T0252105

The Social Neuroscience
of Human–Animal Interaction

The Social Neuroscience of Human–Animal Interaction

◆

EDITED BY

Lisa S. Freund, Sandra McCune, Layla Esposito,
Nancy R. Gee, and Peggy McCardle

AMERICAN PSYCHOLOGICAL ASSOCIATION

WASHINGTON, DC

Published by
American Psychological Association
750 First Street, NE
Washington, DC 20002
www.apa.org

To order
APA Order Department
P.O. Box 92984
Washington, DC 20090-2984
Tel: (800) 374-2721; Direct: (202) 336-5510
Fax: (202) 336-5502; TDD/TTY: (202) 336-6123
Online: www.apa.org/pubs/books
E-mail: order@apa.org

In the U.K., Europe, Africa, and the Middle East, copies may be ordered from
American Psychological Association
3 Henrietta Street
Covent Garden, London
WC2E 8LU England

Typeset in Goudy by Circle Graphics, Inc., Columbia, MD

Printer: United Book Press, Baltimore, MD
Cover Designer: Berg Design, Albany, NY

The opinions and statements published are the responsibility of the authors, and such opinions and statements do not necessarily represent the policies of the American Psychological Association.

Library of Congress Cataloging-in-Publication Data

Names: Freund, Lisa S., editor. | McCune, Sandra, 1963- editor. | Esposito, Layla, editor. | Gee, Nancy R., editor. | McCardle, Peggy D., editor.
Title: The social neuroscience of human-animal interaction / [edited by] Lisa S. Freund, Sandra McCune, Layla Esposito, Nancy R. Gee, and Peggy McCardle.
Description: Washington, DC : American Psychological Association, [2016] | Includes bibliographical references and index.
Identifiers: LCCN 2015033073 | ISBN 9781433821769 | ISBN 1433821761
Subjects: LCSH: Human-animal relationships. | Animals—Therapeutic use. | Neuropsychology.
Classification: LCC QL85 .S626 2016 | DDC 590—dc23 LC record available at http://lccn.loc.gov/2015033073

British Library Cataloguing-in-Publication Data
A CIP record is available from the British Library.

Printed in the United States of America
First Edition

http://dx.doi.org/10.1037/14856-000

CONTENTS

CONTRIBUTORS

Karen Bales, PhD, Department of Psychology, University of California, Davis

Andrea Beetz, PhD, Department of Special Education, University of Rostock, Rostock, Germany, and Department of Behavioral Biology, University of Vienna, Vienna, Austria

Kristin M. Brethel-Haurwitz, BS, Department of Psychology, Georgetown University, Washington, DC

Casey Brown, PhD, Department of Psychology, University of California, Berkeley

C. Sue Carter, PhD, Kinsey Institute, Indiana University, Bloomington

James A. Coan, PhD, Department of Psychology, University of Virginia, Charlottesville

Adele Diamond, PhD, Division of Developmental Cognitive Neuroscience, Department of Psychiatry, University of British Columbia, Vancouver, Canada

Nancy A. Dreschel, DVM, PhD, Department of Animal Science, Pennsylvania State University, University Park

Layla Esposito, PhD, Child Development and Behavior Branch, *Eunice Kennedy Shriver* National Institute of Child Health and Human Development, National Institutes of Health, Bethesda, MD

Nathan A. Fox, PhD, Department of Human Development and Quantitative Methodology, University of Maryland, College Park

Lisa S. Freund, PhD, Child Development and Behavior Branch, *Eunice Kennedy Shriver* National Institute of Child Health and Human Development, National Institutes of Health, Bethesda, MD

Nancy R. Gee, PhD, Waltham Centre for Pet Nutrition, Leicestershire, United Kingdom

Douglas A. Granger, PhD, Institute for Interdisciplinary Salivary Bioscience Research, Arizona State University, Tempe; Johns Hopkins University School of Nursing, Johns Hopkins University Bloomberg School of Public Health, and Johns Hopkins University School of Medicine, Baltimore, MD

Kun Guo, PhD, School of Psychology, University of Lincoln, Brayford Pool, United Kingdom

Karyl J. Hurley, DVM, DACVIM, Global Scientific Affairs, Mars Incorporated, West Chester, PA

Paul Jones, PhD, Waltham Centre for Pet Nutrition, Leicestershire, United Kingdom

Melissa Kelly, PhD, Division of Developmental Cognitive Neuroscience, Department of Psychiatry, University of British Columbia, Vancouver, Canada

Kurt Kotrschal, PhD, Department of Behavioral Biology and Konrad Lorenz Forschungsstelle, University of Vienna, Vienna, Austria

Daphne S. Ling, BS, Division of Developmental Cognitive Neuroscience, Department of Psychiatry, University of British Columbia, Vancouver, Canada

Leah M. Lozier, PhD, Department of Neurology, Georgetown University, Washington, DC

Abigail A. Marsh, PhD, Department of Psychology, Georgetown University, Washington, DC

Peggy McCardle, PhD, MPH, Haskins Laboratories, New Haven, CT; Peggy McCardle Consulting, LLC, Seminole, FL, and Annapolis, MD

Sandra McCune, PhD, Waltham Centre for Pet Nutrition, Leicestershire, United Kingdom

Stephen W. Porges, PhD, Kinsey Institute, Indiana University, Bloomington

Paul C. Quinn, PhD, Department of Psychological and Brain Sciences, University of Delaware, Newark

John M. Rawlings, PhD, Waltham Centre for Pet Nutrition, Leicestershire, United Kingdom

The Social Neuroscience
of Human–Animal Interaction

INTRODUCTION

NANCY R. GEE, LAYLA ESPOSITO, SANDRA McCUNE,
LISA S. FREUND, AND PEGGY McCARDLE

Humans are a fundamentally social species, preferring to live in dyads, families, groups, communities, and cultures (Cacioppo & Ortigue, 2011). As a species, we have a wonderful capacity to develop and engage in social interactions, both with other humans and with members of other species, most obviously companion animals such as dogs and cats. In fact, the latest figures indicate that 68% of U.S. households (American Pet Products Association, 2014; American Veterinary Medicine Association, 2007) and 46% of British households (Pet Food Manufacturers Association, 2014) include at least one companion animal. Pet ownership has also been shown to facilitate "social capital" in that the presence of pets tends to facilitate social contact and a sense of community (Wood, Giles-Corti, & Bulsara, 2005; Wood et al., 2015).

The views expressed in this volume are those of the authors and do not necessarily represent those of the National Institutes of Health, the *Eunice Kennedy Shriver* National Institute of Child Health and Human Development, the U.S. Department of Health and Human Services, or Mars, Incorporated.

http://dx.doi.org/10.1037/14856-001
The Social Neuroscience of Human–Animal Interaction, L. S. Freund, S. McCune, L. Esposito, N. R. Gee, and P. McCardle (Editors)

A number of studies have suggested that human–animal interaction (HAI) may play a role in improving human health (Anderson, Reid, & Jennings, 1992; Friedmann & Thomas, 1995; Friedmann, Thomas, & Eddy, 2000; Gorrity & Stallones, 1998), preventing emotional distress, reducing stress (Allen, Blascovich, Tomaka, & Kelsey, 1991; Serpell, 1991), and increasing well-being across the lifespan. The multidisciplinary field of HAI is still young, with much of its research base being descriptive in nature (relying on survey methods and correlational techniques) and with little focus on specifics such as the underlying neurobiological mechanisms associated with HAI.

THE PROMISE OF SOCIAL NEUROSCIENCE

Like HAI, social neuroscience is also a relatively new interdisciplinary field, emerging over the past 2 decades. Cacioppo and Berntson (1992) popularized the term *social neuroscience*, which incorporates a multilevel approach to the study of mental and behavioral phenomena. Social neuroscience is an interdisciplinary academic field devoted to understanding how biological systems implement social processes and behavior, and how these social structures and processes impact the brain and biology. Historically, this field grew out of two disciplines that represented previously separate lines of scientific inquiry. One focused on the investigation of the human experience of interacting with one another and how those interactions, real or imagined, influence thought, feelings, and behavior. The other, grounded in biology and biomedicine, focused on understanding how our biological makeup enables such things as behavior, cognition, and emotion, as well as how to explain, predict, and ultimately prevent such things as illness, disease, and psychological disorders. By bridging the gap between these two lines of inquiry, social neuroscience contributes to our understanding of the mechanisms by which our social interactions impact health, lifespan, and cognition (Cacioppo & Ortigue, 2011).

HAI PROVIDES A RICH AREA FOR SOCIAL NEUROSCIENCE INVESTIGATION

HAI is a fertile area for social neuroscientific investigations into our neural, genetic, and basic affiliation and motivation processes. Because this area of public health inquiry has only recently emerged, researchers with expertise in studying neurobiological mechanisms supporting human behaviors have rarely been involved in HAI and are not necessarily familiar with the intriguing public health results and questions generated by early HAI research. Likewise, researchers who have been involved in HAI studies are typically less familiar with neurobiological approaches or the types of collaborations they could establish to investigate basic biological mechanisms underlying the effects of HAI.

ORIGIN OF THIS BOOK

In 2008, the *Eunice Kennedy Shriver* National Institute of Child Health and Human Development (NICHD) and the WALTHAM Centre for Pet Nutrition, a division of Mars, Inc., entered a public–private partnership to explore HAI, in particular, as it pertains to child development and health. To further explore the possibilities, the partners held a series of meetings that brought together researchers and practitioners working in animal behavior and cognition, and child health and development, with those studying or clinically applying principles of HAI. Out of these meetings grew two books. The first (McCardle, McCune, Griffin, & Maholmes, 2011) was targeted to research psychologists seeking to know more about research design and methods as used in behavioral investigations of HAI in homes and community settings as well as in intervention studies. The second (McCardle, McCune, Griffin, Esposito, & Freund, 2011) focused more on the human side of HAI, on the need for evidence-based practice in a field that is growing rapidly. In 2011, a workshop was held on the social neuroscience of HAI, which forms the basis of this volume, addressing the basic neurobiological mechanisms that underlie the effects observed in HAI.

This volume brings together researchers who study the neural bases of attachment, social bonds, emotions in social interactions, cognition and the social bases of cognition, responses to social stressors, and genetics. Some of these researchers are investigating HAI; others primarily study humans but have interests in HAI. The volume raises relevant questions regarding basic mechanisms of HAI of mutual interest to human and animal behaviorists and neurobiological researchers, as well as questions about the methods and measures appropriate for investigating those questions.

DEVELOPMENTS IN HAI AND SOCIAL NEUROSCIENCE

The chapters and integrative commentaries in this volume demonstrate that interdisciplinary social neuroscience investigations within and across species are becoming increasingly valuable. Contributors represent multiple disciplines, with a preponderance of psychologists but also biologists, ethologists, psychiatrists, veterinarians, and an ethicist. Their research is conducted in countries throughout North America and Europe.

An area of particular interest to social neuroscience is the effects of neuropeptides on humans, nonhuman animals, and the relationships between them (see especially the chapters in Part II). Research in this area has advanced our understanding of the relationship between pituitary hormones and both pair bonding and social behaviors, and it provides an exciting direction for researchers exploring the nature of both social and biological aspects of attachment and bonding in human–companion animal relationships.

SCOPE AND ORGANIZATION OF THIS BOOK

The volume is based on three main themes: Cognition fundamental to social neuroscience; applying neuroscience to HAI; and science and research considerations, that is, methods and approaches to help to unlock the potential of research in this field. It presents predominantly research involving dogs and horses, as these are the animals with which the most HAI research has been conducted. But as the research agenda makes clear, research on HAI and animal-assisted interventions should include all animals currently involved in these activities, and the important focus on the mechanisms of action that underlie the results being reported for HAI and related interventions is critical to the field's progress and to our more complete understanding of the relationship between humans and animals.

In the first part of this volume, *Cognition: Setting the Stage for Deeper Social Neuroscience*, psychologists Quinn and Guo each discuss research related to perceptual recognition and identification of stimulus cues. Quinn (Chapter 1) presents his work on cognitive categorization, specifically, how young infants mentally organize information from stimuli around them, including people, places, and things, including cats and dogs. Guo (Chapter 2) discusses visual attention and recognition of facial cues in both humans and animals. The ability to categorize stimuli (including animals) and the ability to recognize and identify facial cues are fundamental aspects of human cognition known to influence social interactions and are likely to influence how humans and animals recognize one another, interact, and potentially form attachments.

Coming from a cognitive neuroscience perspective, Ling, Kelly, and Diamond (Chapter 3) discuss the development of executive functions in children and offer indications of how HAI may link to this set of developmental abilities that are crucial to academic success and overall success in life. In making this connection between the development of executive functions and HAI, they lay the foundation for the possibility that the presence of animals in the lives of children has the potential to enhance learning environments and help them to achieve a greater level of success in life than if they had not been exposed to animals during their early development.

This first part concludes with an integrative commentary by biologist Kotrschal, who has worked extensively on the topic of HAI and attachment. Based on his experience in examining both the animal and the human side of the HAI equation, while also focusing on the importance of the formation of attachment to social development, his thoughtful commentary provides a unique and valuable perspective to help set the stage for linking work on social cognition and neuroscience to the investigation of mechanisms underlying the effects of HAI.

In the second part, *Neurobiology: Applying Neuroscience to Human–Animal Interaction*, Carter and Porges (Chapter 4), a behavioral neurobiologist and a psychobiologist, respectively, delve into the evolutionary development of the autonomic nervous system and the neuropeptides oxytocin and vasopressin as they apply to HAI. Psychologists Beetz and Bales (Chapter 5) then explore how these molecules may affect affiliation and attachment in HAI. Brown and Coan (Chapter 6) outline their research on the social regulation of humans' response to threat, including their social baseline theory, and how this social ecology could apply to humans and their household pets. In Chapter 7, Lozier, Brethel-Haurwitz, and Marsh discuss the relationship between empathy, psychopathy, and animal abuse, viewed through the lens of cognitive neuroscience.

This part also concludes with a thought-provoking integrative commentary. Freund, a developmental psychologist and cognitive neuroscientist, highlights and extends the possibilities these authors raise for a deeper exploration of the social neuroscience of HAI to enable a mechanistic account of HAI's effects.

In the third part, *Science and Research Considerations*, geneticist Jones and ethologist McCune (Chapter 8), after a presentation of information about the genetic components of companion animals, describe recent advances in genetic technologies and discuss how genetic diversity in both dogs and cats can be clearly linked to specific behaviors. This chapter also highlights the promise for future research in behavioral genetics to inform HAI, such as in the selection of the best-suited animals for therapeutic interventions. In Chapter 9, veterinarian and biobehaviorist Dreschel and psychologist Granger detail the applications and limitations of salivary bioscience. Their discussion includes both human and animal bioscience considerations, including saliva sample collection, research design, and analytical strategies, as well as specific analytes in saliva that are likely to be of interest to research on HAI. Because the sampling of oral fluid is minimally invasive, and the number of substances that can be reliably measured using this approach is increasing, salivary bioscience has the potential to make profound contributions to the future of research and theory in the social neuroscience of HAI.

In the final chapter (Chapter 10) in this part, psychologist Gee, veterinarian Hurley, and ethicist Rawlings team up to address the animal welfare implications of HAI research and practice. The authors focus specifically on dogs as the animal most commonly involved in animal-assisted interventions. They describe behavioral expectations for the dog and provide specific recommendations for dog selection, training, and evaluation. Most important from a welfare perspective, they describe behavioral indicators of stress in dogs, make recommendations for reducing that stress, and, finally, discuss the unique challenges of working with special populations. In the integrative commentary for this part, McCardle highlights the value of these chapters, encouraging both researchers interested in HAI and those seeking to better understand the complex science that will of necessity be involved in some social neuroscience of HAI research to further their own education in these key areas. She also emphasizes the importance and value of interdisciplinary collaborations if we are to take advantage of the recent scientific advances made in other fields.

CONCLUSION

Over the past several decades, a considerable amount of evidence has amassed demonstrating that elements in the psychological and physiological domains can be influenced by a multiplicity of factors (Plomin, 1989). Cacioppo and Berntson (1992) made the case that more can be achieved by considering a multilevel integrative approach that jointly pursues the understanding of phenomena at both a micro level (biological/neurological) and a macro level (social psychology). We concur with this view and see it as a natural step in the exploration of the underlying mechanisms that can in part explain the—until now—seemingly inexplicable attachment humans have to their pets, the comfort they draw from them, and the already observed but largely unexplained health, developmental, and psychological benefits such interactions apparently provide.

REFERENCES

Allen, K. M., Blascovich, J., Tomaka, J., & Kelsey, R. M. (1991). Presence of human friends and pet dogs as moderators of autonomic responses to stress in women. *Journal of Personality and Social Psychology, 61,* 582–589. http://dx.doi.org/10.1037/0022-3514.61.4.582

American Pet Products Association. (2014). *Pet industry market size & ownership statistics.* Retrieved from http://www.americanpetproducts.org/press_industry-trends.asp

American Veterinary Medicine Association. (2007). *U.S. pet ownership and demographics sourcebook*. Schaumburg, IL: Author.

Anderson, W. P., Reid, C. M., & Jennings, G. L. (1992). Pet ownership and risk factors for cardiovascular disease. *The Medical Journal of Australia, 157*, 298–301.

Cacioppo, J. T., & Berntson, G. G. (1992). Social psychological contributions to the decade of the brain: Doctrine of multilevel analysis. *American Psychologist, 47*, 1019–1028. http://dx.doi.org/10.1037/0003-066X.47.8.1019

Cacioppo, J. T., & Ortigue, S. (2011). Social neuroscience: How a multidisciplinary field is uncovering the biology of human interactions. *Cerebrum, 17*.

Friedmann, E., & Thomas, S. A. (1995). Pet ownership, social support, and one-year survival after acute myocardial infarction in the Cardiac Arrhythmia Suppression Trial (CAST). *The American Journal of Cardiology, 76*, 1213–1217. http://dx.doi.org/10.1016/S0002-9149(99)80343-9

Friedmann, E., Thomas, S. A., & Eddy, T. J. (2000). Companion animals and human health: Physical and cardiovascular influences. In A. L. Podberscek, E. Paul, & J. Serpell (Eds.), *Companion animals and us: Exploring the relationship between people and pets* (pp. 125–142). New York, NY: Cambridge University Press.

Gorrity, T. F., & Stallones, L. (1998). Effects of pet contact on human well-being: Review of recent research. In C. C. Wilson & D. C. Turner (Eds.), *Companion animals in human health* (pp. 3–22). Thousand Oaks, CA: Sage. http://dx.doi.org/10.4135/9781452232959.n1

McCardle, P., McCune, S., Griffin, J. A., Esposito, L., & Freund, L. S. (Eds.). (2011). *Animals in our lives: Human–animal interaction in family, community, and therapeutic settings*. Baltimore, MD: Paul H. Brookes.

McCardle, P., McCune, S., Griffin, J. A., & Maholmes, V. (Eds.). (2011). *How animals affect us: Examining the influence of human–animal interaction on child development and human health*. Washington, DC: American Psychological Association.

Pet Food Manufacturers Association. (2014). *Statistics*. Retrieved from http://www.pfma.org.uk/statistics/

Plomin, R. (1989). Environment and genes. Determinants of behavior. *American Psychologist, 44*, 105–111. http://dx.doi.org/10.1037/0003-066X.44.2.105

Serpell, J. (1991). Beneficial effects of pet ownership on some aspects of human health and behaviour. *Journal of the Royal Society of Medicine, 84*, 717–720.

Wood, L., Giles-Corti, B., & Bulsara, M. (2005). The pet connection: Pets as a conduit for social capital? *Social Science & Medicine, 61*, 1159–1173. http://dx.doi.org/10.1016/j.socscimed.2005.01.017

Wood, L., Martin, K., Christian, H., Nathan, A., Lauritsen, C., Houghton, S., . . . McCune, S. (2015). The pet factor—companion animals as a conduit for getting to know people, friendship formation and social support. *PLoS ONE, 10*(4), e0122085. http://dx.doi.org/10.1371/journal.pone.0122085

I

COGNITION: SETTING THE STAGE FOR DEEPER SOCIAL NEUROSCIENCE

1

WHAT DO INFANTS KNOW ABOUT CATS, DOGS, AND PEOPLE? DEVELOPMENT OF A "LIKE-PEOPLE" REPRESENTATION FOR NONHUMAN ANIMALS

PAUL C. QUINN

Infancy is an important time window for learning how knowledge is acquired. During the first months of life, infants experience objects from various classes, but at different rates of presentation. For instance, infants are likely to encounter other humans more frequently than nonhuman animals. This aspect of infants' early experience allows investigators to examine whether the category representations that infants form for humans versus nonhuman animals differ in terms of their exclusivity and structure. Researchers can also inquire about the perceptual information and type of processing (i.e., bottom-up vs. top-down) used to categorize humans and nonhuman animals. The present chapter reviews convergent evidence from looking time, computational, and brain-imaging methods indicating that the contrast between how infants respond to humans versus nonhuman animals

Preparation of this chapter was supported by Grant R01 HD-46526 from the *Eunice Kennedy Shriver* National Institute of Child Health and Human Development. The author thanks the editors for their helpful comments on an earlier draft.

http://dx.doi.org/10.1037/14856-002
The Social Neuroscience of Human–Animal Interaction, L. S. Freund, S. McCune, L. Esposito, N. R. Gee, and P. McCardle (Editors)

exemplifies an expert–novice difference in the early development of category representations. This difference suggests that infants may respond to nonhuman animals in a human-centric manner and use a "like-people" foundation for a commonsense biology. An implication is that persons, even very early in development, may form strong attachments with nonhuman animals because humans perceive nonhuman animals to be like other humans.

THE MENTAL PROCESS OF CATEGORIZATION

The central cognitive process at issue in the chapter is that of *categorization*, which refers to the recognition of different entities as members of the same category based on some internal representation of the category (Murphy, 2002). Categorization is considered to be a critical cognitive ability, with some arguing that it is the most fundamental activity to be carried out by the nervous system (Edelman, 1987) and the most foundational of all cognitive processes (Thelen & Smith, 1994). This is because a system of mental representation lacking in category representation would be dominated by unrelated instance information, and individuals would have to learn to respond anew to each novel object encountered. That is, without categories, each new entity encountered in the world would be unrelated to entities previously encountered. It is via categorization processes that memory storage emerges as organized, retrieval of information becomes efficient, and individuals become capable of responding with familiarity to an indefinitely large number of exemplars from multiple categories, a majority of which will not have been previously experienced.

HISTORICAL VIEWS ON THE DEVELOPMENTAL ORIGINS OF CATEGORIZATION

Because of the importance of categorization as a mental process that supports everyday cognitive functioning, there has been interest in understanding how it begins and is manifested during early development. Different theoretical accounts have been offered. An associative learning view is that categories are formed through pairings of environmental encounters and verbal labels (Hull, 1920). By this view, a young child sees a dog and hears it labeled as such. Over time, as the experience is repeated, the child will arrive at a meaning for the concept of "dogs." An anthropological view is that the world has little or no natural order: The child must learn through instruction to impose order on natural disorder (Leach, 1964).

Although there is little doubt that some of our skill at dividing the world into categories is derived from language and formal instruction, with both mechanisms still the subject of contemporary inquiry (e.g., Csibra & Gergely, 2009; Dean, Kendal, Schapiro, Thierry, & Laland, 2012; Rakison & Lupyan, 2008), the question arises as to whether something even more fundamental might underlie our ability to categorize, especially in the initial months of life, when the contributions of language and instruction toward concept formation may be minimal (but see Ferry, Hespos, & Waxman, 2010). In response to this question, Rosch (1978) argued that the world is structured along lines of natural discontinuity, such that categories are marked by bundles of correlated attributes. For example, birds tend to have feathers, two legs, and make chirping sounds, whereas dogs tend to have fur, four legs, and make barking sounds. By this view, an organism that can detect such correlations and compile them into separate representations is capable of categorization, and if that is the case, then at least the initial start-up of categorization may begin without the benefit of language and instruction. It therefore becomes important to understand the categorization abilities of infants, as it may be from these abilities that the more complex categories of children and adults are derived.

A METHOD TO STUDY CATEGORIZATION IN YOUNG INFANTS

To study categorization in young infants (approximately 3–7 months of age), researchers have adapted a looking time methodology based on novelty preference (Fantz, 1964). The basic idea is that when it comes to generic objects, infants will look longer at the novel relative to the familiar. In studies of categorization, infants are familiarized with instances from a common category (e.g., different instances of cats). Then, during a novel category preference test, two new instances are presented. One is a novel instance from the familiarized category (a cat not previously seen), and the other is a novel instance from a novel category (e.g., a bird or dog). If the infant generalizes responsiveness to new instances of the familiarized category (i.e., looks for about the same amount of time to the new cat relative to the cats shown during familiarization) and responds differentially to novel instances from the novel category (i.e., displays a visual preference for a novel bird or dog), then it may be concluded that the infants have formed a category representation for the class of familiarized cat exemplars that includes novel instances from the familiarized category (i.e., novel cats) and excludes novel instances from the novel category (i.e., novel birds or dogs). This conclusion is also contingent on the results of control conditions demonstrating that infants can discriminate the instances from the familiarized category and that the preference for the novel category exemplars did not arise simply from an a priori preference.

Experimental Evidence

Given that some of the earliest words in a child's expressive vocabulary include *kitty, dog, bird,* and *horse* (Gentner, 1982; Gleason, 2005; Nelson, 1973), it seemed a natural step to investigate how young infants represent nonhuman animal categories (Eimas & Quinn, 1994; Quinn & Eimas, 1998; Quinn, Eimas, & Rosenkrantz, 1993). Infants were presented with a dozen images of cats or horses during familiarization. Those infants familiarized with cats were administered novel category preference tests pairing novel cat images with novel images of birds, dogs, horses, and people, whereas those infants familiarized with horses were administered novel category preference tests pairing novel horse images with novel images of cats, giraffes, zebras, and people.

Examples of the categories can be seen in Figure 1.1. The images from the nonhuman animal categories included different breeds, coloring, and stances, and the images of humans included both males and females in standing, running, or walking poses. Moreover, it can be seen that size differences between the categories were minimized. Had size not been removed as a cue, and infants differentiated between the categories, then the studies would

Figure 1.1. Examples of the nonhuman animal and human images presented in the studies investigating infants' categorization of nonhuman animals.

have been vulnerable to the critique that only simple size discrimination had been demonstrated. It is also important to note that each category consisted of 18 different images. Such large stimulus sets allowed each infant to be familiarized and tested with randomly selected images from the target and contrast categories.

The findings were that infants familiarized with cats generalized responsiveness to novel cats and displayed novel category preferences for birds, dogs, horses, and people. In addition, infants familiarized with horses generalized responsiveness to novel horses and displayed novel category preferences for cats, giraffes, zebras, and people. Control studies showed that the results of the novel category preference tests could not be attributed to within-category discrimination failure (which would have rendered the various instances of the familiarized category indistinguishable from one another, thus making the novel category preference test more like a simple discrimination test) or to a priori preferences for the contrast categories. These outcomes indicate that infants can form category representations for nonhuman animal categories that include novel instances of those categories but exclude members of contrasting nonhuman animal categories as well as instances of humans (for corroborating results, see also Mareschal, Powell, & Volein, 2003; Oakes & Ribar, 2005; Younger & Fearing, 1999).

Why Is Categorization by Infants Significant?

The results of the studies investigating categorization of nonhuman animals by infants are significant because they suggest that infants have the ability to divide the world of objects into perceptual category representations that later come to have conceptual significance for children and adults. As such, category development may reflect enrichment of category representations initially formed through perceptual experience (Madole & Oakes, 1999; Quinn & Eimas, 1997; Rakison & Poulin-Dubois, 2001). More concretely, the evidence indicates that infants are not representing the world as an undifferentiated bunch of grapes. Rather, when presented with realistic, pictorial instances of the categories, infants form distinct representations of, for example, cats and dogs. These representations may serve as placeholders for the acquisition of the more nonobvious, essentialist attributes that occurs beyond infancy. That is, children learn and adults know that cats have cat DNA and give birth to kittens, whereas dogs have dog DNA and give birth to puppies (Gelman, 2003; Keil, 1989). It may be the process of incorporating these more abstract attributes into the category representations of infants that provides a pathway to mature concepts. Indeed, it's difficult to envision how one could go from the undifferentiated bunch of grapes starting point to the mature concept end point without the perceptual category representations of infants serving as a facilitating intermediary. It should also be acknowledged that what is being attributed to infant performance is not that infants are learning long-term memory-based concepts of cats and dogs with which they leave the lab. The claim is that infants are demonstrating perceptual parsing skills in the laboratory that can presumably be deployed when infants encounter real cats and dogs in the natural environment.

A Role for Part Information

What is the perceptual information that infants use to form category representations for nonhuman animals? In some cases, such as when horses are contrasted with zebras, distinctive body markings (e.g., the stripes of the zebras) may provide the basis for the category partitioning. However, in other cases, such as the contrasts between cats, dogs, and horses, the diagnostic information is not obvious, given that each species possesses a head, a body, four legs, fur, and a tail. To investigate the perceptual information that infants use to categorize cats versus dogs, the categorization experiment contrasting cats and dogs was repeated. However, in this instance, only information from the heads (bodies occluded) or bodies (heads occluded) was presented during familiarization and the novel category preference test (Quinn & Eimas, 1996). Infants displayed a novel category preference in the head-only, but not the body-only, condition, suggesting that the head is used to partition the categories. This conclusion was bolstered by an additional experimental outcome from a procedure in which infants were familiarized with cat and dog images and then tested with hybrid images, that is, dog head on cat body and cat head on dog body (Spencer, Quinn, Johnson, & Karmiloff-Smith, 1997). The novel category preference was found to follow the direction of the novel category head. Both results suggest that the category representations of infants for nonhuman animals are in some instances based on part information (see Rakison & Butterworth, 1998, for further evidence on the importance of parts as a basis for category formation by infants).

A Summary Representation

A further question concerns the nature of the representations that infants form for nonhuman animal categories. Are infants simply storing individual instances of the category in accord with an exemplar-based representation (Nosofsky, 2011), or are they extracting just the summary information about the category in accord with more of a prototype representation (Minda & Smith, 2011)? A prototype is the central value extracted from a series of varying exemplars (e.g., Homa, Cross, Cornell, Goldman, & Schwartz, 1973; Posner & Keele, 1968). To investigate this issue, the familiarization portion of the categorization experiment in which infants were presented with a dozen images of cats or horses was repeated. However, in this instance, the ensuing preference test contrasted a novel image from the familiarized category with one of the images presented during familiarization (randomly selected for each infant). If infants are representing the individual instances of category, then they should display a preference for the novel instance of the category. If, however, they are representing only summary information, then one would not expect a preference between the novel and familiarized exemplars. The outcome was no preference toward the novel over the familiarized (i.e., a null preference), which is in accord with the idea that the category representations of infants for nonhuman animals are based on summary information, perhaps in the form of individual prototypes corresponding to each category (e.g., Bomba & Siqueland, 1983; Quinn, 1987). This finding suggests that infants are representing "cat" rather than individual cats in the case of cats as the familiarized category. Although caution should be exercised in terms of drawing conclusions on the basis of no preference between instances, the results take on additional significance when contrasted with the different finding observed when humans are the familiarized category (to be discussed subsequently).

A Role for Online Learning Taking Place During Category Familiarization

Another question concerning the category representations formed by infants for nonhuman animals is whether those representations are formed online during the course of the experimental familiarization procedure or whether the experimental procedure is tapping into a representation that existed prior to arrival at the laboratory. One source of evidence relevant to this question comes from computational learning systems that are trained with measurements of the surface attributes of the cat and dog images, the same images that were presented to infants in the laboratory (e.g., Mareschal, French, & Quinn, 2000). Such learning systems were able to simulate infant performance in their category-learning behavior. This result suggests that the category learning can be based on bottom-up processing of the image statistics, which in turn is consistent with the idea that the category representations of infants for nonhuman animal categories are formed online during the course of the experimental familiarization procedure. This conclusion is also consistent with the observation that at least some of the infant participants in the categorization studies did not have cats or dogs as pets at home and were not likely to have directly experienced the target category of horses or the contrast categories of giraffes or zebras prior to participation in the studies. A possible facilitative influence of a home pet such as a cat or dog on categorization performance will be discussed in a subsequent section.

A second piece of evidence comes from measurement of event-related potentials (ERPs) in infants in an ERP analogue of the behavioral categorization task (Quinn, Westerlund, & Nelson, 2006). Infants were presented with 18 cat images for 500 ms per image during the first half of familiarization, another 18 images during the second half of familiarization, then 20 new cat images intermixed with 20 dog images. If infants are learning the cat images during familiarization, then the response to the second set of cats should differ from the response to the first set of cats because it will reflect a learned category structure. The response to the learned category structure should then generalize to the novel cats, but not to the novel dogs. The response to the novel dogs should look more like the response to the initial set of cats, reflecting in both instances initial experience with exemplars of a category.

In studies of simple recognition memory in infants using ERP methods, a late slow-wave component that peaks between 1,000 and 1,500 ms post-stimulus is associated with recognition of familiarity and detection of novelty (de Haan & Nelson, 1997). In particular, if the wave returns to baseline, then that is associated with recognition of familiarity, whereas if the wave deflects away from baseline in a negative direction (a component called the *negative slow wave*), that is associated with detection of novelty. For the late slow-wave activity recorded over left occipital-parietal scalp, responses to the initial set of familiarized cats and novel dogs deflected away from baseline as a negative slow wave, reflecting recognition of their novelty. In contrast, responses to the second set of familiarized cats and the novel cats returned to baseline, reflecting recognition of familiarity. The ERP results are thus consistent with the computational results in suggesting that the cat and dog images are learned online during the course of experience with the images.

CATEGORIZATION OF HUMANS BY INFANTS

Experimental Evidence

Given positive evidence of categorization of nonhuman animals by infants and mindful of the perceptual differences between nonhuman animals and humans, one might expect that it would be straightforward to demonstrate that infants would form a category representation of humans inclusive of novel humans, but exclusive of nonhuman animals. To this end, infants were familiarized with a dozen images of humans and then administered novel category preference tests in which novel humans were paired with cats, horses, fish, and even cars (Quinn & Eimas, 1998). Surprisingly, the infants generalized responsiveness not only to the novel humans, but also to the cats, horses, and fish, although they additionally displayed a novel category preference for the cars. It is important to note that control studies showed that infants did not display a spontaneous preference for humans over nonhuman animals, thus ruling out an a priori preference as a possible basis for the generalized responsiveness to the nonhuman animal contrast categories. The infants thus seemed to form a broadly inclusive representation for humans that included novel humans as well as non-human animals but excluded cars. Pauen (2000) also reported evidence for a broadly inclusive representation of humans by infants that includes non-human animals.

What Is the Basis for the Asymmetry?

Why would infants include nonhuman animals in their representation for humans but exclude contrasting nonhuman animal categories and humans from their representation of nonhuman animals? That is, why would infants include cats, horses, and fish in their representation of humans, but exclude birds, dogs, horses, and humans from their representation of cats, and exclude cats, zebras, giraffes, and humans from their representation of horses? A consideration is that in contrast to learning about the nonhuman animal images online during the course of an experiment, even the youngest infants in the studies at issue have acquired several months of experience with humans prior to participation in the studies. This consideration leads to the suggestion that infants may be at a more advanced stage of processing images of humans relative to images of nonhuman animals, which in turn gives rise to the proposal that what infants know about nonhuman animals may be likened to what novices know about generic object categories, whereas what infants know about humans may be likened to what experts know about objects in their domain of expertise.

To evaluate the proposal further, prior studies have pointed to at least three advantages that experts display over novices when perceiving objects within their domain of expertise. First, experts are better at recognizing commonalities among objects within their domain of expertise (Murphy & Wright, 1984). Notably, this characteristic of expertise is already evident in infant responding to the equivalence of humans, cats, horses, and fish. It could be that infants are recognizing the abstract resemblance of a head, attached to an elongated body, with skeletal appendages across all four categories. The recognition of such a resemblance may in turn be the basis for the human representation being an attractor for nonhuman animals. Second, experts tend to recognize objects within their domain of expertise at a subordinate level. That is, instead of recognizing a feathered animal at the generic level of bird as a novice would, an expert would recognize the object as a "sparrow" or "white-crowned sparrow" (Gauthier & Tarr, 1997; Tanaka & Taylor, 1991). Third, experts tend to recognize objects within their domain of expertise as configural wholes rather than on the basis of individual parts (Busey & Vanderkolk, 2005; Gauthier & Tarr, 2002). The next two sections consider whether there is evidence that infants represent humans at the subordinate level and as organized wholes.

An Exemplar-Based Representation

As was asked about infant representation of nonhuman animal categories, one can inquire about the structure of the representation infants use to form a category for humans (Quinn & Eimas, 1998). The data indicate that infants represent the human category at the level of specific exemplars. In particular, when familiarized with a dozen humans and tested with one of the familiar humans (randomly selected) and a novel human, infants prefer the novel human. This result stands in contrast to the null result that was observed when infants were administered this same type of familiarization procedure and preference test, but with cats or horses as stimuli. The findings with the human stimuli are consistent with the idea that infants recognize humans at an expert, subordinate level.

A Role for the Whole

As indicated earlier, when infants form category representations for nonhuman animals that exclude contrasting nonhuman animal categories (i.e., a representation for cats that excludes dogs), they do so on the basis of perceptual part information (i.e., the head). However, when the whole stimulus versus head only (body occluded) versus body only (head occluded) categorization study is repeated with cats and humans (instead of cats and dogs), the major finding is that it's only in the whole stimulus condition that cats are included in the human representation (Quinn, 2004). Moreover, a follow-up study showed that the inclusion of the cats into the representation for humans occurs only when the whole stimulus images are upright, suggesting a role for configural processing in the formation of a representation for humans that includes nonhuman animals (Quinn, Lee, Pascalis, & Slater, 2007). Taken together, these studies indicate that infants include nonhuman animals in their representation for humans based on holistic-configural stimulus information, which is in accord with the idea that expert-like processing underlies the recognition of humans by infants.

A Contribution From Previously Acquired Knowledge

As noted, computational and ERP evidence suggest that infant learning of nonhuman animals takes place primarily via online learning. By contrast, there is computational evidence suggesting that infant responding to human images is based on learning about humans that occurred prior to arrival at the laboratory (Mermillod, French, Quinn, & Mareschal, 2004). For example, if the same neural network that simulated infant learning of cat and dog images learns about humans and nonhuman animals just from the bottom-up information available in the stimulus images, the network does not produce the categorization asymmetry observed in the infant's looking time behavior in the laboratory. However, if the network is provided with a long-term memory that was trained with images of humans, then the network does produce the asymmetry (i.e., a representation for humans that includes nonhuman animals and representations for nonhuman animals that exclude humans). This result suggests that prior knowledge about humans contributes to the formation of a category representation for humans that includes nonhuman animals.

Further support for the idea that how infants respond to humans in the laboratory is based on prior knowledge comes from studies on how infants respond to social category attributes of humans, namely, the gender and race of human faces (Quinn et al., 2013). In particular, processing advantages are observed for the categories that infants experience more frequently on a day-to-day basis. For example, infants raised by female primary caregivers represent female faces as individual exemplars but represent male faces as a summary structure (Quinn, Yahr, Kuhn, Slater, & Pascalis, 2002). In addition, infants come to display superior recognition abilities for faces from within one's own racial group (Kelly et al., 2007, 2009).

SUMMARY

The evidence reviewed has documented differences in the way that infants represent nonhuman animals versus humans. Specifically, infants represent nonhuman animal categories as exclusive of both humans and contrasting nonhuman animal categories, on the basis of part information, with a summary representation and because of learning taking place at the laboratory within the experimental task. In contrast, infants represent humans as inclusive of nonhuman animals, on the basis of holistic-configural information, with an exemplar representation, and by recruiting from a preexisting knowledge base about humans.

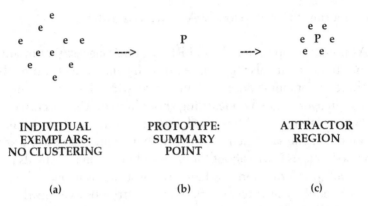

INDIVIDUAL EXEMPLARS: NO CLUSTERING	PROTOTYPE: SUMMARY POINT	ATTRACTOR REGION
(a)	(b)	(c)

Figure 1.2. Schematic depiction of the course of early knowledge acquisition.

An Accounting of the Differences

The differences in how infants represent humans versus nonhuman animals may be quantitative in nature and reflect different levels of knowledge that infants have about humans versus nonhuman animals. Figure 1.2 provides a schematic illustration depicting a possible course of early knowledge acquisition in which infants start out experiencing individual exemplars from multiple categories. With sufficient experience and armed with inherent abilities to compute within-category similarity and between-category dissimilarity, infants begin to form structured representations for these categories based on summary-level prototypes. The expectation is that such representations would be formed for a variety of generic object categories, including those for nonhuman animals. It is this type of representation that reflects the way a novice represents generic object category information. More generally, the argument is that core processes extracting the similarity structure of the environment may largely account for the representation of such categories.

With additional experience that comes in the form of large numbers of exemplars, or rich or lengthy interaction with a smaller number of exemplars, the summary representation is supplemented with information about individual instances, and it is in this process that the representation comes to take on an expert nature and acquires its attractor or magnet-like properties. More broadly, the claims are that differences in the frequency of exposure to different classes are needed to account for how infants represent humans, and what comes first may play an important role in structuring the perceptual field. The significance of such attractor representations for the larger course of cognitive development comes in their potential for organizing large portions of experience. That is, the inclusion of nonhuman animals into the representation of humans may be an important step in the development of a broad, domain-level representation of animals in general. By this view, humans may be the glue that provides the coherence for the concept of animal.

Further Thoughts on the Longer Term Developmental Significance of the Infant's Representation for Humans

The incorporation of nonhuman animals into a representation by infants may remind some readers of the findings of Carey (1985), who reported that children reason about nonhuman animals on the basis of their similarity to a human prototype. Moreover, Inagaki and Hatano (1999, 2006), while discussing the Carey findings and related results of their own, attributed the outcomes to the more substantial knowledge that children have about humans relative to nonhuman animals. The argument here is that the seeds of the human-based inference or person analogy system of reasoning in children may be planted very early in development and may be rooted in infants' differential experience with humans versus nonhuman animals.

Does This Mean Humans Are Special?

The thesis advanced in the present chapter should not be construed as a "humans are special" argument. For example, there was nothing special about the human category that was stored in the long-term memory component of the computational network described earlier (Mermillod et al., 2004). This observation leads to the prediction that any category to which infants or children are exposed repeatedly could serve as an attractor for categories perceived to have some commonalities. In support of this idea is the finding that children who have experience raising goldfish use goldfish as well as humans as a basis for their projections about other animals in children's inductive reasoning tasks (Inagaki, 1990).

If humans are not special and pets such as goldfish can provide organizing information acquired about other animals, then one may ask whether infants with a cat or dog at home display expert-like processing advantages for the corresponding categories. The evidence on this question has been mixed, with initial reports suggesting no effect but subsequent studies that used more of an individual differences approach suggesting that there may in fact be an effect. Consistent with the argument advanced earlier in the chapter that learning about cats and dogs in the laboratory is based predominantly on online learning processes, first reports were that infants presented with a cat or dog category-learning task and with a cat or dog at home (that was a match with the cat or dog category in the task) did not show advantages in preferential responding to a novel contrast category at either the basic or subordinate level (Quinn, 2004, 2010). These findings should perhaps not be viewed as surprising given that the infants experienced only a single exemplar of the category (the cat or dog pet) and because the richness, frequency, and duration of an infant's encounters with a pet would not be expected to match that experienced with humans. Moreover, anecdotal reports from parents who express concern over infants falling victim to the bite of a pet dog or scratch of a pet cat and who take steps to control and even limit infant–pet interactions are consistent with this expectation. An additional consideration when interpreting the null outcomes is that they may reflect to some extent a ceiling on novel category preferences. Such preferences tend to range between chance responding, at 50%, to above-chance novel category preference scores, in the range from 60% to 65%. Given that infants without pets already categorize nonhuman animal images at above-chance levels, that does not leave much room for a facilitative effect of a home pet to be measured on top of the already above-chance performance achieved without a home pet.

Other observations are consistent with the idea that a home pet can facilitate infant categorization performance, and some subsequent studies have provided support for these observations. In particular, single exemplars have been shown to have facilitative effects on category representation and induction tasks in the case of the caregiver's influence on the gender category in infancy (Quinn, Yahr, Kuhn, Slater, & Pascalis, 2002) and the pet goldfish's influence on nonhuman animal categories in childhood (Inagaki, 1990). Moreover, Kovack-Lesh, Horst, and Oakes (2008) reported that experience with pets facilitated infants' categorization of cats and dogs. Notably, the effect was found only among a subset of the infants living in those households (i.e., those who actively compared the exemplars presented during familiarization) and was independent of whether the category learned during the laboratory familiarization session matched with the category of pet present in the household, although a follow-up study suggested that the effect is stronger when the cat or dog category learned in the laboratory matches with the type of pet at home (Kovack-Lesh, Oakes, & McMurray, 2012).

It will be interesting to determine in future work whether additional expert processing advantages (i.e., configural processing, attractor effects) accrue for infants being reared with a cat or dog in their household. Computational evidence suggests that at least in the case of attractor effects, the answer will be yes, as Rogers and McClelland (2004) reported that neural networks trained on dogs generalize broadly from dogs to goats and robins, although not to trees. One possible caveat on the prediction is that the networks were trained on multiple dogs (not just single exemplars). It could be that such category learning is in part what serves to tune perceptual systems to features that will be useful to learning subsequently presented categories (Quinn & Tanaka, 2007) and gives rise to this type of expertise effect.

CONCLUSION

One may ask how the account advanced in this chapter connects with the broader themes presented in the volume as a whole. One theme of emphasis concerns social neuroscience, and recent evidence has begun to identify possible neural correlates that may be associated with the inclusion of nonhuman animals into a category representation of humans by infants (Marinović, Hoehl, & Pauen, 2014). Another theme of emphasis concerns the mechanisms by which humans develop companion relationships with nonhuman animals such as cats and dogs. The present chapter suggests that such relationships may develop in part because nonhuman animals are perceived to be like humans. It has been argued that infants and young children may compute similarity in relation to internal reference points to guide their responding in various domains of development. For example, infants may rely on a "like-me" representation when imitating the behavior of others (Meltzoff, 2007) and also rely on a "like-caregiver" representation when looking at the faces of others (Quinn et al., 2010). The results reviewed here suggest that infants and young children may use a "like-people" representation when developing their common-sense notions of animals and the biological world.

REFERENCES

Bomba, P. C., & Siqueland, E. R. (1983). The nature and structure of infant form categories. *Journal of Experimental Child Psychology, 35,* 294–328. http://dx.doi.org/10.1016/0022-0965(83)90085-1

Busey, T. A., & Vanderkolk, J. R. (2005). Behavioral and electrophysiological evidence for configural processing in fingerprint experts. *Vision Research, 45,* 431–448. http://dx.doi.org/10.1016/j.visres.2004.08.021

Carey, S. (1985). *Conceptual change in childhood.* Cambridge, MA: MIT Press.

Csibra, G., & Gergely, G. (2009). Natural pedagogy. *Trends in Cognitive Sciences, 13,* 148–153. http://dx.doi.org/10.1016/j.tics.2009.01.005

Dean, L. G., Kendal, R. L., Schapiro, S. J., Thierry, B., & Laland, K. N. (2012). Identification of the social and cognitive processes underlying human cumulative culture. *Science, 335,* 1114–1118. http://dx.doi.org/10.1126/science. 1213969

de Haan, M., & Nelson, C. A. (1997). Recognition of the mother's face by six-month-old infants: A neurobehavioral study. *Child Development, 68,* 187–210. http://dx.doi.org/10.2307/1131845

Edelman, G. M. (1987). *Neural Darwinism: The theory of neuronal group selection.* New York, NY: Basic Books.

Eimas, P. D., & Quinn, P. C. (1994). Studies on the formation of perceptually based basic-level categories in young infants. *Child Development, 65,* 903–917. http://dx.doi.org/10.2307/1131427

Fantz, R. L. (1964). Visual experience in infants: Decreased attention to familiar patterns relative to novel ones. *Science, 146,* 668–670. http://dx.doi.org/10.1126/science.146.3644.668

Ferry, A., Hespos, S. J., & Waxman, S. (2010). Categorization in 3- and 4-month-old infants: An advantage of words over tones. *Child Development, 81,* 472–479. http://dx.doi.org/10.1111/j.1467-8624.2009.01408.x

Gauthier, I., & Tarr, M. J. (1997). Becoming a "Greeble" expert: Exploring mechanisms for face recognition. *Vision Research, 37,* 1673–1682. http://dx.doi.org/10.1016/S0042-6989(96)00286-6

Gauthier, I., & Tarr, M. J. (2002). Unraveling mechanisms for expert object recognition: Bridging brain activity and behavior. *Journal of Experimental Psychology: Human Perception and Performance, 28,* 431–446. http://dx.doi.org/10.1037/0096-1523.28.2.431

Gelman, S. A. (2003). *The essential child: Origins of essentialism in everyday thought.* New York, NY: Oxford University Press. http://dx.doi.org/10.1093/acprof:oso/9780195154061.001.0001

Gentner, D. (1982). Why nouns are learned before verbs: Linguistic relativity versus natural partitioning. In S. Kuczaj (Ed.), *Language development, Vol. 2: Language, thought, and culture* (pp. 301–334). Hillsdale, NJ: Erlbaum.

Gleason, J. B. (2005). *The development of language* (6th ed.). Boston, MA: Allyn & Bacon.

Homa, D., Cross, J., Cornell, D., Goldman, D., & Schwartz, S. (1973). Prototype abstraction and classification of new instances as a function of number of instances defining the prototype. *Journal of Experimental Psychology, 101,* 116–122. http://dx.doi.org/10.1037/h0035772

Hull, C. L. (1920). Quantitative aspects of the evolution of concepts. *Psychological Monographs, 28*(1, Whole No. 123), 1–86.

Inagaki, K. (1990). The effects of raising animals on children's biological knowledge. *British Journal of Developmental Psychology, 8,* 119–129. http://dx.doi.org/10.1111/j.2044-835X.1990.tb00827.x

Inagaki, K., & Hatano, G. (1999). Children's understanding of mind-body relationships. In M. Siegal & C. Peterson (Eds.), *Children's understanding of biology and health* (pp. 23–44). Cambridge, England: Cambridge University Press. http://dx.doi.org/10.1017/CBO9780511659881.003

Inagaki, K., & Hatano, G. (2006). Young children's conception of the biological world. *Current Directions in Psychological Science, 15,* 177–181. http://dx.doi.org/10.1111/j.1467-8721.2006.00431.x

Keil, F. C. (1989). *Concepts, kinds, and cognitive development.* Cambridge, MA: MIT Press.

Kelly, D. J., Liu, S., Lee, K., Quinn, P. C., Pascalis, O., Slater, A. M., & Ge, L. (2009). Development of the other-race effect during infancy: Evidence toward universality? *Journal of Experimental Child Psychology, 104,* 105–114. http://dx.doi.org/10.1016/j.jecp.2009.01.006

Kelly, D. J., Quinn, P. C., Slater, A. M., Lee, K., Ge, L., & Pascalis, O. (2007). The other-race effect develops during infancy: Evidence of perceptual narrowing. *Psychological Science, 18,* 1084–1089. http://dx.doi.org/10.1111/j.1467-9280.2007.02029.x

Kovack-Lesh, K. A., Horst, J. S., & Oakes, L. M. (2008). The cat is out of the bag: The joint influence of previous experience and looking behavior on infant categorization. *Infancy, 13,* 285–307. http://dx.doi.org/10.1080/15250000802189428

Kovack-Lesh, K. A., Oakes, L. M., & McMurray, B. (2012). Contributions of attentional style and previous experience to 4-month-old infants' categorization. *Infancy, 17,* 324–338. http://dx.doi.org/10.1111/j.1532-7078.2011.00073.x

Leach, E. (1964). Anthropological aspects of language: Animal categories and verbal abuse. In E. H. Lenneberg (Ed.), *New directions in the study of language* (pp. 23–63). Cambridge, MA: MIT Press.

Madole, K. L., & Oakes, L. M. (1999). Making sense of infant categorization: Stable processes and changing representations. *Developmental Review, 19,* 263–296. http://dx.doi.org/10.1006/drev.1998.0481

Mareschal, D., French, R. M., & Quinn, P. C. (2000). A connectionist account of asymmetric category learning in early infancy. *Developmental Psychology, 36,* 635–645. http://dx.doi.org/10.1037/0012-1649.36.5.635

Mareschal, D., Powell, D., & Volein, A. (2003). Basic-level category discriminations by 7- and 9-month-olds in an object examination task. *Journal of Experimental Child Psychology, 86,* 87–107. http://dx.doi.org/10.1016/S0022-0965(03)00107-3

Marinović, V., Hoehl, S., & Pauen, S. (2014). Neural correlates of human-animal distinction: An ERP-study on early categorical differentiation with 4- and 7-month-old infants and adults. *Neuropsychologia, 60,* 60–76. http://dx.doi.org/10.1016/j.neuropsychologia.2014.05.013

Meltzoff, A. N. (2007). 'Like me': A foundation for social cognition. *Developmental Science, 10,* 126–134. http://dx.doi.org/10.1111/j.1467-7687.2007.00574.x

Mermillod, M., French, R. M., Quinn, P. C., & Mareschal, D. (2004). The importance of long-term memory in infant perceptual categorization. In R. Alterman

& D. Kirsh (Eds.), *Proceedings of the 25th Annual Conference of the Cognitive Science Society* (pp. 804–809). Mahwah, NJ: Erlbaum.

Minda, J. P., & Smith, J. D. (2011). Prototype models of categorization: Basic formulation, predictions, and limitations. In E. Pothos & A. Wills (Eds.), *Formal approaches in categorization* (pp. 40–64). Cambridge, England: Cambridge University Press. http://dx.doi.org/10.1017/CBO9780511921322.003

Murphy, G. L. (2002). *The big book of concepts*. Cambridge, MA: MIT Press.

Murphy, G. L., & Wright, J. C. (1984). Changes in conceptual structure with expertise: Differences between real-world experts and novices. *Journal of Experimental Psychology: Learning, Memory, and Cognition, 10,* 144–155. http://dx.doi.org/10.1037/0278-7393.10.1.144

Nelson, K. (1973). Structure and strategy in learning to talk. *Monographs of the Society for Research in Child Development, 38*(1-2, Serial No. 149), 1–137.

Nosofsky, R. M. (2011). The generalized context model: An exemplar model of classification. In E. Pothos & A. Wills (Eds.), *Formal approaches in categorization* (pp. 18–39). Cambridge, England: Cambridge University Press. http://dx.doi.org/10.1017/CBO9780511921322.002

Oakes, L. M., & Ribar, R. J. (2005). A comparison of infants' categorization in paired and successive familiarization tasks. *Infancy, 7,* 85–98. http://dx.doi.org/10.1207/s15327078in0701_7

Pauen, S. (2000). Early differentiation within the animate domain: Are humans something special? *Journal of Experimental Child Psychology, 75,* 134–151. http://dx.doi.org/10.1006/jecp.1999.2530

Posner, M. I., & Keele, S. W. (1968). On the genesis of abstract ideas. *Journal of Experimental Psychology, 77,* 353–363. http://dx.doi.org/10.1037/h0025953

Quinn, P. C. (1987). The categorical representation of visual pattern information by young infants. *Cognition, 27,* 145–179. http://dx.doi.org/10.1016/0010-0277(87)90017-5

Quinn, P. C. (2004). Development of subordinate-level categorization in 3- to 7-month-old infants. *Child Development, 75,* 886–899. http://dx.doi.org/10.1111/j.1467-8624.2004.00712.x

Quinn, P. C. (2010). The acquisition of expertise as a model for the growth of cognitive structure. In S. P. Johnson (Ed.), *Neoconstructivism: The new science of cognitive development* (pp. 252–273). New York, NY: Oxford University Press.

Quinn, P. C., Anzures, G., Lee, K., Pascalis, O., Slater, A., & Tanaka, J. W. (2013). On the developmental origins of differential responding to social category information. In M. R. Banaji & S. A. Gelman (Eds.), *Navigating the social world: What infants, children, and other species can teach us* (pp. 286–291). New York, NY: Oxford University Press. http://dx.doi.org/10.1093/acprof:oso/9780199890712.003.0052

Quinn, P. C., Conforto, A., Lee, K., O'Toole, A. J., Pascalis, O., & Slater, A. M. (2010). Infant preference for individual women's faces extends to girl prototype faces. *Infant Behavior & Development, 33,* 357–360. http://dx.doi.org/10.1016/j.infbeh.2010.03.001

Quinn, P. C., & Eimas, P. D. (1996). Perceptual cues that permit categorical differentiation of animal species by infants. *Journal of Experimental Child Psychology, 63,* 189–211. http://dx.doi.org/10.1006/jecp.1996.0047

Quinn, P. C., & Eimas, P. D. (1997). A reexamination of the perceptual-to-conceptual shift in mental representations. *Review of General Psychology, 1,* 271–287. http://dx.doi.org/10.1037/1089-2680.1.3.271

Quinn, P. C., & Eimas, P. D. (1998). Evidence for a global categorical representation of humans by young infants. *Journal of Experimental Child Psychology, 69,* 151–174. http://dx.doi.org/10.1006/jecp.1998.2443

Quinn, P. C., Eimas, P. D., & Rosenkrantz, S. L. (1993). Evidence for representations of perceptually similar natural categories by 3-month-old and 4-month-old infants. *Perception, 22,* 463–475. http://dx.doi.org/10.1068/p220463

Quinn, P. C., Lee, K., Pascalis, O., & Slater, A. M. (2007). In support of an expert-novice difference in the representation of humans versus non-human animals by infants: Generalization from persons to cats occurs only with upright whole images [Special issue]. *Cognition, Brain, & Behavior, 11,* 679–694.

Quinn, P. C., & Tanaka, J. W. (2007). Early development of perceptual expertise: Within-basic-level categorization experience facilitates the formation of subordinate-level category representations in 6- to 7-month-old infants. *Memory & Cognition, 35,* 1422–1431. http://dx.doi.org/10.3758/BF03193612

Quinn, P. C., Westerlund, A., & Nelson, C. A. (2006). Neural markers of categorization in 6-month-old infants. *Psychological Science, 17,* 59–66. http://dx.doi.org/10.1111/j.1467-9280.2005.01665.x

Quinn, P. C., Yahr, J., Kuhn, A., Slater, A. M., & Pascalis, O. (2002). Representation of the gender of human faces by infants: A preference for female. *Perception, 31,* 1109–1121. http://dx.doi.org/10.1068/p3331

Rakison, D. H., & Butterworth, G. E. (1998). Infants' use of object parts in early categorization. *Developmental Psychology, 34,* 49–62. http://dx.doi.org/10.1037/0012-1649.34.1.49

Rakison, D. H., & Lupyan, G. (2008). Developing object concepts in infancy: An associative learning perspective. *Monographs of the Society for Research in Child Development, 73,* 1–110.

Rakison, D. H., & Poulin-Dubois, D. (2001). Developmental origin of the animate-inanimate distinction. *Psychological Bulletin, 127,* 209–228. http://dx.doi.org/10.1037/0033-2909.127.2.209

Rogers, T. T., & McClelland, J. L. (2004). *Semantic cognition: A parallel distributed processing approach.* Cambridge, MA: MIT Press.

Rosch, E. (1978). Principles of categorization. In E. Rosch & B. B. Lloyd (Eds.), *Cognition and categorization* (pp. 27–48). Hillsdale, NJ: Erlbaum.

Spencer, J., Quinn, P. C., Johnson, M. H., & Karmiloff-Smith, A. (1997). Heads you win, tails you lose: Evidence for young infants categorizing mammals by head and facial attributes (Special issue: Perceptual Development). *Early Development & Parenting, 6,* 113–126. http://dx.doi.org/10.1002/(SICI)1099-0917(199709/12)6:3/4<113::AID-EDP151>3.0.CO;2-W

Tanaka, J. W., & Taylor, M. (1991). Object categories and expertise: Is the basic level in the eye of the beholder? *Cognitive Psychology, 23,* 457–482. http://dx.doi.org/10.1016/0010-0285(91)90016-H

Thelen, E., & Smith, L. B. (1994). *A dynamic systems approach to the development of cognition and action.* Cambridge, MA: MIT Press.

Younger, B. A., & Fearing, D. D. (1999). Parsing items into separate categories: Developmental change in infant categorization. *Child Development, 70,* 291–303. http://dx.doi.org/10.1111/1467-8624.00022

2

VISUAL ATTENTION AND FACIAL IDENTIFICATION IN HUMAN AND NONHUMAN ANIMALS

KUN GUO

One important research topic in social neuroscience and human–animal interaction (HAI) is to understand how facial identification contributes to interspecies understanding. The perception of a face not only provides visual information about an individual's gender, age, and familiarity but also can communicate significant cues to affective state (e.g., happy), intention (e.g., gaze direction), cognitive activity (e.g., concentration), and temperament (e.g., hostility). The ability to recognize these cues and to respond promptly and appropriately guides effective human social interactions (Bruce & Young, 2012).

The capacity to read faces is not restricted to humans and can be found extensively in nonhuman primates and other social species. In the phylogeny of primates, there is a trend toward larger and more complex social groups in which individuals rely more on visual cues (e.g., facial signals) than on olfactory cues for communication. Numerous studies have demonstrated that nonhuman primates, such as monkeys and chimpanzees, possess visual behaviors

http://dx.doi.org/10.1037/14856-003
The Social Neuroscience of Human–Animal Interaction, L. S. Freund, S. McCune, L. Esposito, N. R. Gee, and P. McCardle (Editors)

and cognitive and cortical mechanisms for face perception similar to those of humans (Emery, 2000; Leopold & Rhodes, 2010). This is not particularly surprising given their close evolutionary relationship to humans and their naturally complex social environments.

Domestic dogs (*Canis familiaris*), on the other hand, are more distantly related to humans phylogenetically. However, probably because of their close social association with humans and enculturation (dogs have been domesticated for at least 10,000 years and possibly much longer; Vilà et al., 1997), they have shown greater attachment (Topál et al., 2005) and attention bias (Miklósi et al., 2003) toward humans compared with their close relative, the wolf. Their sensitivity to human communication cues exceeds that of nonhuman primates in certain tasks such as following human gaze direction, and it is hypothesized that they may have evolved a special predisposition for communicating with humans that may represent a case of convergent cognitive evolution (Hare & Tomasello, 2005). Also, given that human–dog interaction is a typical model in studying HAI, it seems essential to understand how domestic dogs process facial cues, especially those of humans. However, because of various technical and ethical challenges, little is known about the neural mechanisms of face perception in dogs.

In this chapter, I mainly use experimental evidence from behavioral and eye-tracking studies to compare the capability of sampling and processing several facial cues (i.e., facial identity, facial expression, and eye gaze) across humans, nonhuman primates, and domestic dogs.

PERCEIVING FACIAL CUES OF IDENTITY, EXPRESSION, AND GAZE DIRECTION

Humans

Although different human faces share the same basic configuration (i.e., two eyes above a nose and mouth), and the same face could have different appearances (influenced by hairstyle, makeup, aging, diet, etc.), humans are highly efficient at differentiating and identifying faces, possibly through a holistic or configural process of perceiving relations among individual facial features and integrating all features into an individual representation of the face as a whole (Bruce & Young, 2012). This configural processing is often assessed by the face-inversion effect, which is defined as a larger decrease in recognition performance for faces than for other single-oriented objects when they are presented upside down (Valentine, 1988). Although the precise cause is still a source of debate (qualitative vs. quantitative differences between the processing of upright and inverted faces; e.g., Sekuler, Gaspar,

Gold, & Bennett, 2004), the general consensus is that this face-inversion effect results mainly from a disruption in the processing of orientation-sensitive facial configural information.

In addition to facial identity, humans also have very high perceptual sensitivities to facial expressions of emotion, especially those representing our typical emotional states such as happiness, sadness, fear, anger, disgust, and surprise, some even with very brief presentation (<100 ms; e.g., Willis & Todorov, 2006) or when focal attention is not fully available (Bruce & Young, 2012).

Another facial cue to which humans are extremely sensitive is gaze direction. Perceived eye contact modulates cognition and attention. For instance, direct gaze facilitates face-related behavioral tasks, such as face detection, gender discrimination, and identity recognition (Senju & Johnson, 2009). Considering the importance of gaze interplay and gaze following in social interaction and learning (Emery, 2000), it is not surprising that even human infants primarily use the orientation of the eyes, rather than the head, to determine another's direction of gaze (Tomasello, Hare, Lehmann, & Call, 2007).

Nonhuman Primates

Like humans, nonhuman primates are also able to perceive differences between individuals of their own species based on facial cues alone. At an early age, monkeys respond appropriately to the expressions of other individuals (Mendelson, Haith, & Goldman-Rakic, 1982) and can recognize their faces (Rosenfeld & Van Hoesen, 1979). More important, they can discriminate face images of unfamiliar individuals, as demonstrated through match-to-sample or visual paired comparison procedures in chimpanzees (*Pantroglodytes*) and monkeys (*Macaca mulatta*, *Macaca tonkeana*, *Cebus paella*; Gothard, Brooks, & Peterson, 2009; Parr, Dove, & Hopkins, 1998). This face discrimination and recognition performance is generally better for faces from their own species than from other species (Dufour, Pascalis, & Petit, 2006). These studies have provided compelling evidence that nonhuman primates can perceive face identity purely based on visual cues presented within black-and-white face images.

There are, however, some subtle differences in face recognition capability between species. Chimpanzees show the face-inversion effect when tested with both upright and inverted conspecific faces (Parr et al., 1998; Tomonaga, 2007), suggesting they may use a human-like configural process. On the other hand, the inverted faces lead to idiosyncratic response strategies and sometimes contradictory results in monkeys, suggesting that if they do use configural processing, it is to a lesser degree than humans and chimpanzees (Leopold & Rhodes, 2010).

Both chimpanzees and monkeys display a wider range of facial expressions in social interactions and can reliably interpret (at least some species-specific) expressions, both live and in photographs (Leopold & Rhodes, 2010). Through practice, chimpanzees can even categorize human and monkey facial expressions (Kanazawa, 1996). Both species also accurately perceive gaze direction from conspecifics, as prolonged direct gaze often serves as an aggressive gesture to reinforce submissive gaze aversion in lower ranking group members (Emery, 2000). They, however, are less sensitive to human gaze direction, but instead tend to follow human head orientation (Leopold & Rhodes, 2010).

Domestic Dogs

Like human and nonhuman primates, dogs also display various facial expressions, such as aggressive (i.e., direct stare, the baring of teeth), submissive, and fearful displays, suggesting that facial communication plays a role in dog–dog and dog–human social interactions (Feddersen-Petersen, 2005). Although dogs' visual acuity (20/50–20/100 with the Snellen chart) is lower than that of primates and they have comparatively less binocular overlap, less range of accommodation, and limited color perception (dichromatic color vision; Miller & Murphy, 1995), their visual system and brain structure do not impose serious limitations on the use of visual cues for social cognition and interaction. In fact, probably because of their unique history of domestication and selective breeding, domestic dogs are unusually skilled, in some cases even more so than nonhuman primates, at reading human social-communicative signals, including various facial cues (Hare & Tomasello, 2005).

Many behavioral studies have shown that dogs exhibit certain sensitivities to human facial cues. When forced to choose between two humans with food, they prefer to beg from the person with visible face and eyes rather than the person with covered eyes or head (Gácsi, Miklósi, Varga, Topál, & Csányi, 2004). They are more likely to avoid approaching forbidden food when a human's eyes are open than when they are closed (Call, Bräuer, Kaminski, & Tomasello, 2003). Further, like human infants but unlike chimpanzees, dogs are sensitive to human gaze cues and can "understand" their communicative meaning. They use human head and eye direction cues to locate hidden food only if the person is gazing directly at one of two possible target locations and ignore it if the gaze is directed above the target (Soproni, Miklósi, Topál, & Csányi, 2001).

A few recent studies have further demonstrated that dogs can process facial cues presented in black-and-white pictures. In a visual paired comparison (preferential looking) procedure, dogs displayed a clear looking preference for a novel human face image when simultaneously presented with a prior-exposed familiar human face; however, the dogs directed longer viewing time at a familiar dog face when paired with a novel dog face, suggesting that they can use species-specific facial cues alone to differentiate individual dogs and humans. Interestingly, no significant looking preference was detected for inverted human or dog face images, implying that dogs may also use a human-like configural strategy in face perception (Racca et al., 2010).

In addition to facial identity information, dogs also show sensitivity for processing other facial cues presented in face pictures, such as discriminating smiling faces from blank facial expressions (Nagasawa, Murai, Mogi, & Kikusui, 2011). However, in comparison with extensive studies in human and nonhuman primates, it is still largely unclear what types of facial information dogs are sensitive to, to what extent they can respond to these facial signals, and what perceptual and neural mechanisms underlie their face perception.

COGNITIVE BIAS AND BRAIN LATERALIZATION IN FACE PROCESSING

Humans

Although facial configuration is more or less symmetrical along the vertical axis, humans are more likely to use facial cues contained in the right side of the owner's face (left side, from the viewer's perspective) to facilitate perceptual judgement of gender, age, identity, expression, likeness, and attractiveness (Burt & Perrett, 1997). For instance, when asked to label the facial expression of a briefly presented chimeric face image, in which the left and right sides of the viewed face differ in facial expressions, viewers tend to base their decision more frequently on the visual input from the hemiface appearing in their left visual field (left hemiface). This left perceptual bias in face perception is often accompanied by a left gaze bias (LGB), defined by the higher probability of first gaze and a higher proportion of viewing time directed at the left hemiface through eye tracking or preferential looking measurements, when free eye movements are allowed in face exploration (Butler et al., 2005). In other words, the left hemiface is often inspected first and/or for longer periods.

The LGB in face exploration is related to neither handedness nor eye dominance (Leonards & Scott-Samuel, 2005). Although human visuospatial attention bias is to the left visual field (Nicholls & Roberts, 2002), and in some cultures a well practiced left-to-right directional scanning bias (i.e., reading in alphabetic languages) may contribute to this gaze asymmetry (Heath, Rouhana, & Ghanem, 2005), it is often argued that a right hemisphere advantage in face processing is the underlying neural mechanism

(Butler et al., 2005). Neurological and brain imaging studies have consistently demonstrated that face perception is preferentially lateralized in the brain's right hemisphere. Compared with the left hemisphere, patients with right hemisphere damage are more likely to be impaired in facial identity and expression recognition, and normal volunteers have greater activation in face-sensitive cortical areas (e.g., fusiform face area, occipital face area, and posterior superior temporal sulcus) within the right hemisphere when viewing faces (Kanwisher, 2000). As the right hemisphere receives visual input from the left visual field, it is plausible that the LGB in face exploration is related to the brain's right hemisphere bias in face processing. This hypothesis is further supported by recent observations that the LGB is most evident in viewing upright faces, but is less or not at all evident in viewing inverted faces (face inversion dramatically impairs the efficiency of normal face processing) and symmetric nonface object or landscape images (Guo, Meints, Hall, Hall, & Mills, 2009; Leonards & Scott-Samuel, 2005).

Like many other aspects of face processing capability, the face-related LGB is an acquired behavior, possibly through the process of experience-dependent gradual specialization during development. For instance, 6-month-old infants show a general, inherent LGB for both face and nonface object images, which later transforms itself into a more specific LGB for faces only, which has been shown in 4-year-old children and adults (Guo et al., 2009; Racca, Guo, Meints, & Mills, 2012), indicating a progressive narrowing or specializing of perceptual ability in face perception. Interestingly, this face-specific gaze bias is not restricted to human faces. Recent studies have observed consistent and indistinguishable LGB when human observers view human, monkey, dog, and cat faces with neutral facial expressions (Guo, Tunnicliffe, & Roebuck, 2010).

As the LGBs can occur at different stages of face viewing, they could be driven by different perceptual mechanisms. The early LGB, indicated by the first gaze allocation, could be initiated by the gist perception of facial configuration in an automatic fashion to direct a viewer's attention to the left hemiface. That is, when a face is initially presented within a viewer's central visual field, the left hemiface is projected to the face-sensitive right hemisphere of the viewer, where its saliency is more readily evaluated, causing an increase in the viewer's attention as necessary. The later or overall LGB, indicated by longer viewing time at the left hemiface, could be further associated with the processing of specific facial cues, given that the human left and right hemifaces can transmit the same type of facial cues in different intensities and/or speeds (e.g., evoked anger is expressed more intensely in our right hemiface; Indersmitten & Gur, 2003).

Using realistic human face photos, Guo, Smith, Powell, and Nicholls (2012) observed a consistent leftward bias for both initial gaze and overall fixation distribution when participants were performing different behavioral tasks, such as free viewing, judging face familiarity, and labeling facial expressions. The stronger leftward bias at the initial stage of face viewing suggests that the gist perception of the facial configuration plays a central role in developing the LGB. The indistinguishable overall LGB across different behavioral tasks suggests that the LGB is not sensitive to the acquiring or processing of specific facial information and may not be perception dependent. In other words, the LGB is an automatic reflection of hemispheric lateralization in face processing and is not necessarily correlated with the perceptual processing of a specific type of facial information, at least in human viewers.

Nonhuman Primates

The face-related LGB also occurs in nonhuman primates. Eye-tracking studies have revealed that laboratory-reared rhesus monkeys showed both initial and overall LGB when free viewing upright human and monkey face images with neutral facial expressions (Guo et al., 2009) or movies of monkey facial displays (Mosher, Zimmerman, & Gothard, 2011). No gaze asymmetry was observed once the face images were inverted (Guo et al., 2009). These findings are not surprising, considering that rhesus monkeys rely heavily on facial cues for social communication with conspecifics or human caregivers and possess a viewing behavior for face perception similar to that of humans (Guo, 2007; Guo, Robertson, Mahmoodi, Tadmor, & Young, 2003). In addition, monkeys show hemispheric asymmetry in face processing, with stronger activation in the brain's right hemisphere, similar to humans (Rossion & Gauthier, 2002). Future studies could directly compare monkey and human observers to examine how close these two species are in cognitive biases and brain lateralization associated with face processing.

Domestic Dogs

Using images of human, monkey, and dog faces with neutral facial expressions, measurements of preferential looking have revealed that dogs demonstrated both initial and overall LGB only for human faces but not for monkey or dog faces nor for inanimate object images (Guo et al., 2009). Such face-specific and species-sensitive gaze biases in dogs may have significant adaptive value and could be linked to dogs' unique evolutionary and ontogenetic history. For pet dogs, the ability to extract information from human faces and respond appropriately could have had a selective advantage during the process of domestication, especially as the emotional content of these faces may be of immediate adaptive behavioral significance.

Although human faces and monkey faces share similar facial configurations, dogs did not show gaze bias toward monkey faces. This may be due to their unfamiliarity or irrelevance compared with human faces, although the differentiating criteria remain to be established. However, a failure to show LGB for dog faces might reflect a reduced need or sensitivity in assessing conspecifics with neutral facial expressions, as nonfacial greetings, including olfactory cues and visual cues of body postures, are perhaps of greater significance in this situation.

To further understand the relations between gaze asymmetry, face perception, and emotion processing, Racca and colleagues (Racca et al., 2012) presented dog and human faces with different emotional valences (negative: threatening or anger; neutral; positive: friendly or happy) to pet dogs, using a preferential looking protocol; they observed a consistent LGB for negative and neutral human facial expressions, but no bias for positive expressions. Perhaps dogs interpret human neutral facial expressions as potentially negative, given their lack of clear approach signals. Pet dogs, however, demonstrated a differential gaze asymmetry for dog faces based on their emotional valence, with no gaze bias for neutral expressions but an LGB for negative expressions and a right gaze bias (RGB) for positive expressions (Racca et al., 2012). Similar asymmetrical behavioral responses for emotional stimuli have also been noticed in tail-wagging (Quaranta, Siniscalchi, & Vallortigara, 2007), where the dogs preferentially wagged their tails to the left when presented with emotionally negative stimuli and to the right with positive stimuli. These observations are consistent with the valence model of cerebral lateralization in emotion processing (Ehrlichman, 1986), with the right hemisphere mainly involved in the processing of negative emotions and the left hemisphere mainly involved in the processing of positive emotions.

In comparison with human and nonhuman primates, it seems that gaze asymmetry in dogs is a reflection of brain lateralization not only in face perception but also in emotion processing. Pet dogs present differential lateralized eye movements depending on the species and the emotional valence of the face viewed. They have demonstrated a clear LGB in viewing human and dog faces with negative emotional valence, a RGB in viewing dog faces with positive emotional valence, and no bias for monkey faces. Such observations imply a broader adaptive value of this natural gaze asymmetry in social species, such as domestic dogs.

GAZE DISTRIBUTION IN FACE PROCESSING

Humans

Human face exploration often involves a series of saccades (rapid eye movements) to direct our visual fixation to local facial regions that are informative or interesting to us. The preferred regions are often inspected earlier and attract more fixations and longer viewing time. As movements of the eyes remap the projection of the visual world onto the retina, there is a strong connection between the control of eye movements and the processing of visual inputs. Our gaze distribution in face viewing, hence, provides a real-time behavioral index of ongoing perceptual and cognitive processing and is reflective of our attention, motivation, and preference (Henderson, 2003).

Although human faces have various internal (e.g., eyes, nose, mouth) and external (e.g., hair, face shape) local facial features, we direct the majority of fixations (often > 85% of total fixations or viewing time within a trial) at key internal facial features in face exploration, with the eyes attracting the highest proportion of fixations and viewing time, followed by the nose and mouth (Guo, 2012, 2013; Guo et al., 2010, 2012). Interestingly, this pattern of disproportionate gazing is commonly observed in many different cognitive tasks, such as free viewing, face learning, identity judgement, and facial expression categorization (e.g., Barton, Radcliffe, Cherkasova, Edelman, & Intriligator, 2006; Guo et al., 2010, 2012; Henderson, Williams, & Falk, 2005; Jack, Blais, Scheepers, Schyns, & Caldara, 2009), suggesting a crucial role of the eyes in transmitting various elements of facial information and possibly a generic "built-in" scanning strategy in our brains for general face processing (Guo, 2007, 2012).

The type of task asked of individuals, on the other hand, quantitatively affects gaze distribution at different facial parts. We tend to direct a higher proportion of fixations at the mouth region during a task requiring the identification of a facial expression than in a free-viewing or face familiarity judgement task (Guo et al., 2012). The facial information (e.g., familiarity, facial expressions) also influences gaze distribution to the eyes, nose, and mouth regions. For example, in the face recall task, participants tend to look more at the eyes in familiar than in unfamiliar faces (Heisz & Shore, 2008), although in the facial expression identification task, they fixate more to the eyes in the sad or angry face but more to the mouth in the happy face (Eisenbarth & Alpers, 2011; Guo, 2012, 2013). Taken together these studies indicate that people do look at the local facial regions most characteristic for each type of facial information, and that fixation plays a crucial role in extracting diagnostic information to process different facial cues.

Interestingly, human nonpet owners show a similar pattern of gaze distribution when passively viewing human, monkey, dog, and cat faces (Guo et al., 2010). Among key local facial features, the eyes always attract the highest proportion of fixations and viewing times, followed by the nose and then the mouth. Only the proportion of fixations directed at the mouth region is species dependent, with relatively fewer fixations at the dog and cat mouth. Given the relevance and importance of the human mouth in transmitting a range of expression cues in social communication, and the high similarity in spatial configuration of human and monkey faces, it is quite possible that during free exploration, we involuntarily direct a substantial amount of attention to human and monkey mouths to evaluate subtle expressions. The fewer gazes to the dog and cat mouths from nonpet owners are probably due to their lack of interest and/or perceptual experience in processing subtle emotion cues from dogs and cats. Nevertheless, it appears that the human spontaneous gaze pattern in face viewing is mainly constrained by general facial configurations, but during the course of exploration, prior knowledge/experience about certain face types could influence the distribution of fixations directed at the mouth region (Guo et al., 2010). This hypothesis could be further investigated by comparing the gaze behavior of pet owners and nonpet owners during human–pet interactions.

Nonhuman Primates

As a major animal model for the study of human visual behavior and visual processing, rhesus monkeys display gaze behavior strikingly similar to humans in the viewing of faces. The methods used to assess primate gaze behavior include the search coil (e.g., a rubber ring with an embedded magnet wire adhered to the eye to measure eye movements through electromagnetic induction) or video-based eye tracking. Using these technologies, previous studies have demonstrated that when viewing images of monkey faces (Dahl, Wallraven, Bülthoff, & Logothetis, 2009; Guo et al., 2003), videos of facial displays (Nahm, Perret, Amaral, & Albright, 1997) or vocalizing conspecifics (Ghazanfar, Nielsen, & Logothetis, 2006), monkeys' gaze is frequently directed to key internal facial features, especially the eyes, rather than being randomly or evenly distributed across the whole face. Similar findings have also been reported for other nonhuman primates; baboons (Kyes & Candland, 1987) and chimpanzees (Kano & Tomonaga, 2010) have demonstrated an exaggerated interest in the eye region of the conspecific face. It seems that this social cognitive ability of selectively attending to facial information contained in the eyes has been well preserved across (at least) primate species during evolution.

Monkeys' gaze preference to the eye region starts at the earliest stage of face viewing. Following a face presentation, the first fixation is often directed toward the location of the eyes, even when the eyes are scrambled so as to be unrecognizable (Guo, 2007). However, the chance of the eye region being the first fixation target is significantly reduced after inverting the face or randomly rearranging its spatial location within the face (the local image properties of the eyes remained unchanged; Guo et al., 2003); this suggests that this disproportionate share of initial fixation toward the eyes is driven predominantly by top-down guidance, probably the prior knowledge of their location within the context of a normal face and their essence in facial communication.

Given that monkey and human faces have comparable facial configurations and share some morphological similarities in the evolution of certain facial expressions (e.g., the relaxed open-mouth face and the bared-teeth display in macaques have been proposed to be homologous with laughter and smiling in humans; Preuschoft, 1992), it is not unexpected that the visual system of monkeys is tuned to the same features in the faces of monkeys and humans. In a free-viewing task, Guo and colleagues (Guo et al., 2003) compared monkeys' gaze behavior in viewing monkey and human faces; they observed indistinguishable patterns of within-face gaze distribution, the same time from onset of trial to first saccade and to local facial features, and similar temporal and spatial characteristics of saccadic eye movements (i.e., distribution of saccade distance and individual fixation duration) across the facial images. Using a different experimental protocol, Dahl et al. (2009) found that monkeys directed similar viewing time at the nose and mouth regions of monkey and human faces but longer viewing time at the eyes in monkey than in human faces, suggesting a species-specific face-viewing pattern that may be guided by perceptual expertise.

Nevertheless, although there are some quantitative differences in detailed gaze allocation for individual facial features across different studies that might be caused by different experimental designs, visual stimuli, recording methodology, and analysis protocols, monkey and human viewers exhibit some striking similarities. They show a qualitatively similar gaze strategy in exploring faces (including conspecific and nonconspecific faces) and extracting local facial cues, implying a homologous cognitive mechanism of visual attention in face processing across primates.

Domestic Dogs

Growing interest in canine cognition has promoted research into the allocation of visual attention in domestic dogs. The techniques commonly used to study this (e.g., preferential looking) have, however, lacked spatial accuracy, permitting only gross judgments of the location of the dog's point of gaze and are limited to a laboratory setting. Recently, researchers have started to adapt video-based eye-tracking systems for use with pet dogs, placing a remote eye tracker close to the dog's head and using a sling (Jacobs, Dell'Osso, Hertle, Acland, & Bennett, 2006) or headrest (Somppi, Törnqvist, Hänninen, Krause, & Vainio, 2012) to restrain head/body movements, or attaching a head-mounted eye tracker on the dog's muzzle to allow a more naturalistic behavior (Williams, Mills, & Guo, 2011). These new systems provide a noninvasive method of assessing dogs' viewing behavior with a higher level of spatial accuracy.

So far, only a few eye-tracking studies have examined dogs' gaze behavior in viewing human and dog faces. Like human and nonhuman primates, in an image free-viewing task, dogs preferred to fixate more at conspecific and human faces than at inanimate objects, more at internal facial features (especially at the eyes) than at external features and background scenes, and more at familiar faces and eyes than the strange ones (Somppi et al., 2012; Somppi, Törnqvist, Hänninen, Krause, & Vainio, 2014). When watching videos of a human actor with changing head/gaze directions, dogs fixated at the human face first and showed stronger gaze-following responses if the actor gazed at and verbally addressed the dog directly (Téglás, Gergely, Kupán, Miklósi, & Topál, 2012). Although lacking detailed analysis about gaze distribution within different facial regions, these studies have demonstrated that eye-tracking techniques can be used for studying dogs' social cognition. Future studies could make comprehensive across-species comparisons to examine how dogs extract diagnostic local facial cues within a face and use these cues to guide or facilitate different behavioral tasks, such as emotion recognition.

CONCLUSION

Social species (e.g., humans, nonhuman primates, domestic dogs) extract a rich amount of social information from the faces of their conspecifics and sometimes nonconspecifics. The many shared features of face processing among these species suggest that face perception may have evolved to suit the needs of complex social communication. Studying the presence or the absence of a human-like face-processing behavior in these species, in light of knowledge of their socioecological constraints, provides information regarding the evolutionary connection and selective pressures leading to the emergence of the face-processing system.

Although domestic dogs are a social species and use facial communication, they are not a primarily visually dominant species. Indeed, dogs are not strictly diurnal or nocturnal species and are considered "visual generalists" (Miller & Murphy, 1995). Their observed capability of discriminating faces based on visual cues alone (i.e., 2D pictures) suggests that face perception is not restricted to animal species with high visual acuity and may have wider evolutionary spread in the animal taxon.

Further, the use of spontaneous behaviors, such as visual preferences and gaze distribution, provides a powerful and relatively simple methodological tool to access animals' cognitive abilities. With the increasing availability of many other noninvasive measures, such as eye tracking, electroencephalography, electromyography, and functional magnetic resonance imaging, it is now possible for researchers to adopt a combined approach to systematically study the role of facial communication in HAI (e.g., human–dog interactions) and its underlying cortical mechanisms.

REFERENCES

Barton, J. J. S., Radcliffe, N., Cherkasova, M. V., Edelman, J., & Intriligator, J. M. (2006). Information processing during face recognition: The effects of familiarity, inversion, and morphing on scanning fixations. *Perception, 35,* 1089–1105. http://dx.doi.org/10.1068/p5547

Bruce, V., & Young, A. W. (2012). *Face perception.* London, England: Psychology Press.

Burt, D. M., & Perrett, D. I. (1997). Perceptual asymmetries in judgements of facial attractiveness, age, gender, speech and expression. *Neuropsychologia, 35,* 685–693. http://dx.doi.org/10.1016/S0028-3932(96)00111-X

Butler, S., Gilchrist, I. D., Burt, D. M., Perrett, D. I., Jones, E., & Harvey, M. (2005). Are the perceptual biases found in chimeric face processing reflected in eye-movement patterns? *Neuropsychologia, 43,* 52–59. http://dx.doi.org/10.1016/j.neuropsychologia.2004.06.005

Call, J., Bräuer, J., Kaminski, J., & Tomasello, M. (2003). Domestic dogs (*Canis familiaris*) are sensitive to the attentional state of humans. *Journal of Comparative Psychology, 117,* 257–263. http://dx.doi.org/10.1037/0735-7036.117.3.257

Dahl, C. D., Wallraven, C., Bülthoff, H. H., & Logothetis, N. K. (2009). Humans and macaques employ similar face-processing strategies. *Current Biology, 19,* 509–513. http://dx.doi.org/10.1016/j.cub.2009.01.061

Dufour, V., Pascalis, O., & Petit, O. (2006). Face processing limitation to own species in primates: A comparative study in brown capuchins, Tonkean macaques and humans. *Behavioural Processes, 73,* 107–113. http://dx.doi.org/10.1016/j.beproc.2006.04.006

Ehrlichman, H. (1986). Hemispheric asymmetry and positive-negative affect. In D. Ottoson (Ed.), *Duality and unity of the brain: Unified functioning and specialization of the hemispheres* (pp. 194–206). London, England: Macmillan.

Eisenbarth, H., & Alpers, G. W. (2011). Happy mouth and sad eyes: Scanning emotional facial expressions. *Emotion, 11,* 860–865. http://dx.doi.org/10.1037/a0022758

Emery, N. J. (2000). The eyes have it: The neuroethology, function and evolution of social gaze. *Neuroscience and Biobehavioral Reviews, 24,* 581–604. http://dx.doi.org/10.1016/S0149-7634(00)00025-7

Feddersen-Petersen, D. U. (2005). Communication in wolves and dogs. In V. M. Beko (Ed.), *Encyclopedia of animal behaviour* (Vol. I, pp. 385–394). Westport, CT: Greenwood.

Gácsi, M., Miklósi, Á., Varga, O., Topál, J., & Csányi, V. (2004). Are readers of our face readers of our minds? Dogs (*Canis familiaris*) show situation-dependent recognition of human's attention. *Animal Cognition, 7*, 144–153. http://dx.doi.org/10.1007/s10071-003-0205-8

Ghazanfar, A. A., Nielsen, K., & Logothetis, N. K. (2006). Eye movements of monkey observers viewing vocalizing conspecifics. *Cognition, 101*, 515–529. http://dx.doi.org/10.1016/j.cognition.2005.12.007

Gothard, K. M., Brooks, K. N., & Peterson, M. A. (2009). Multiple perceptual strategies used by macaque monkeys for face recognition. *Animal Cognition, 12*, 155–167. http://dx.doi.org/10.1007/s10071-008-0179-7

Guo, K. (2007). Initial fixation placement in face images is driven by top-down guidance. *Experimental Brain Research, 181*, 673–677. http://dx.doi.org/10.1007/s00221-007-1038-5

Guo, K. (2012). Holistic gaze strategy to categorize facial expression of varying intensities. *PLoS ONE, 7*(8), e42585. http://dx.doi.org/10.1371/journal.pone.0042585

Guo, K. (2013). Size-invariant facial expression categorization and associated gaze allocation within social interaction space. *Perception, 42*, 1027–1042. http://dx.doi.org/10.1068/p7552

Guo, K., Meints, K., Hall, C., Hall, S., & Mills, D. (2009). Left gaze bias in humans, rhesus monkeys and domestic dogs. *Animal Cognition, 12*, 409–418. http://dx.doi.org/10.1007/s10071-008-0199-3

Guo, K., Robertson, R. G., Mahmoodi, S., Tadmor, Y., & Young, M. P. (2003). How do monkeys view faces?—A study of eye movements. *Experimental Brain Research, 150*, 363–374.

Guo, K., Smith, C., Powell, K., & Nicholls, K. (2012). Consistent left gaze bias in processing different facial cues. *Psychological Research, 76*, 263–269. http://dx.doi.org/10.1007/s00426-011-0340-9

Guo, K., Tunnicliffe, D., & Roebuck, H. (2010). Human spontaneous gaze patterns in viewing of faces of different species. *Perception, 39*, 533–542. http://dx.doi.org/10.1068/p6517

Hare, B., & Tomasello, M. (2005). Human-like social skills in dogs? *Trends in Cognitive Sciences, 9*, 439–444. http://dx.doi.org/10.1016/j.tics.2005.07.003

Heath, R. L., Rouhana, A., & Ghanem, D. A. (2005). Asymmetric bias in perception of facial affect among Roman and Arabic script readers. *Laterality: Asymmetries of Body, Brain and Cognition, 10*, 51–64. http://dx.doi.org/10.1080/13576500342000293

Heisz, J. J., & Shore, D. I. (2008). More efficient scanning for familiar faces. *Journal of Vision, 8*, 1–10.

Henderson, J. M. (2003). Human gaze control during real-world scene perception. *Trends in Cognitive Sciences, 7*, 498–504. http://dx.doi.org/10.1016/j.tics.2003.09.006

Henderson, J. M., Williams, C. C., & Falk, R. J. (2005). Eye movements are functional during face learning. *Memory & Cognition, 33*, 98–106. http://dx.doi.org/10.3758/BF03195300

Indersmitten, T., & Gur, R. C. (2003). Emotion processing in chimeric faces: Hemispheric asymmetries in expression and recognition of emotions. *The Journal of Neuroscience, 23,* 3820–3825.

Jack, R. E., Blais, C., Scheepers, C., Schyns, P. G., & Caldara, R. (2009). Cultural confusions show that facial expressions are not universal. *Current Biology, 19,* 1543–1548. http://dx.doi.org/10.1016/j.cub.2009.07.051

Jacobs, J. B., Dell'Osso, L. F., Hertle, R. W., Acland, G. M., & Bennett, J. (2006). Eye movement recordings as an effectiveness indicator of gene therapy in RPE65-deficient canines: Implications for the ocular motor system. *Investigative Ophthalmology & Visual Science, 47,* 2865–2875. http://dx.doi.org/10.1167/iovs.05-1233

Kanazawa, S. (1996). Recognition of facial expressions in a Japanese monkey (*Macaca fuscata*) and humans (*Homo sapiens*). *Primates, 37,* 25–38. http://dx.doi.org/10.1007/BF02382917

Kano, F., & Tomonaga, M. (2010). Face scanning in chimpanzees and humans: Continuity and discontinuity. *Animal Behaviour, 79,* 227–235. http://dx.doi.org/10.1016/j.anbehav.2009.11.003

Kanwisher, N. (2000). Domain specificity in face perception. *Nature Neuroscience, 3,* 759–763. http://dx.doi.org/10.1038/77664

Kyes, R. C., & Candland, D. K. (1987). Baboon (*Papio hamadryas*) visual preferences for regions of the face. *Journal of Comparative Psychology, 101,* 345–348. http://dx.doi.org/10.1037/0735-7036.101.4.345

Leonards, U., & Scott-Samuel, N. E. (2005). Idiosyncratic initiation of saccadic face exploration in humans. *Vision Research, 45,* 2677–2684. http://dx.doi.org/10.1016/j.visres.2005.03.009

Leopold, D. A., & Rhodes, G. (2010). A comparative view of face perception. *Journal of Comparative Psychology, 124,* 233–251. http://dx.doi.org/10.1037/a0019460

Mendelson, M. J., Haith, M. M., & Goldman-Rakic, P. S. (1982). Face scanning and responsiveness to social cues in infant rhesus monkeys. *Developmental Psychology, 18,* 222–228. http://dx.doi.org/10.1037/0012-1649.18.2.222

Miklósi, Á., Kubinyi, E., Topál, J., Gácsi, M., Virányi, Z., & Csányi, V. (2003). A simple reason for a big difference: Wolves do not look back at humans, but dogs do. *Current Biology, 13,* 763–766. http://dx.doi.org/10.1016/S0960-9822(03)00263-X

Miller, P. E., & Murphy, C. J. (1995). Vision in dogs. *Journal of the American Veterinary Medical Association, 207,* 1623–1634.

Mosher, C. P., Zimmerman, P. E., & Gothard, K. M. (2011). Videos of conspecifics elicit interactive looking patterns and facial expressions in monkeys. *Behavioral Neuroscience, 125,* 639–652. http://dx.doi.org/10.1037/a0024264

Nagasawa, M., Murai, K., Mogi, K., & Kikusui, T. (2011). Dogs can discriminate human smiling faces from blank expressions. *Animal Cognition, 14,* 525–533. http://dx.doi.org/10.1007/s10071-011-0386-5

Nahm, F. K. D., Perret, A., Amaral, D. G., & Albright, T. D. (1997). How do monkeys look at faces? *Journal of Cognitive Neuroscience, 9*, 611–623. http://dx.doi.org/10.1162/jocn.1997.9.5.611

Nicholls, M. E. R., & Roberts, G. R. (2002). Can free-viewing perceptual asymmetries be explained by scanning, pre-motor or attentional biases? *Cortex, 38*, 113–136. http://dx.doi.org/10.1016/S0010-9452(08)70645-2

Parr, L. A., Dove, T., & Hopkins, W. D. (1998). Why faces may be special: Evidence of the inversion effect in chimpanzees. *Journal of Cognitive Neuroscience, 10*, 615–622. http://dx.doi.org/10.1162/089892998563013

Preuschoft, S. (1992). "Laughter" and "smile" in Barbary macaques (*Macaca sylvanus*). *Ethology, 91*, 220–236. http://dx.doi.org/10.1111/j.1439-0310.1992.tb00864.x

Quaranta, A., Siniscalchi, M., & Vallortigara, G. (2007). Asymmetric tail-wagging responses by dogs to different emotive stimuli. *Current Biology, 17*, R199–R201. http://dx.doi.org/10.1016/j.cub.2007.02.008

Racca, A., Amadei, E., Ligout, S., Guo, K., Meints, K., & Mills, D. (2010). Discrimination of human and dog faces and inversion responses in domestic dogs (*Canis familiaris*). *Animal Cognition, 13*, 525–533. http://dx.doi.org/10.1007/s10071-009-0303-3

Racca, A., Guo, K., Meints, K., & Mills, D. S. (2012). Reading faces: Differential lateral gaze bias in processing canine and human facial expressions in dogs and 4-year-old children. *PLoS ONE, 7*(4), e36076. http://dx.doi.org/10.1371/journal.pone.0036076

Rosenfeld, S. A., & Van Hoesen, G. W. (1979). Face recognition in the rhesus monkey. *Neuropsychologia, 17*, 503–509. http://dx.doi.org/10.1016/0028-3932(79)90057-5

Rossion, B., & Gauthier, I. (2002). How does the brain process upright and inverted faces? *Behavioral and Cognitive Neuroscience Reviews, 1*, 63–75. http://dx.doi.org/10.1177/1534582302001001004

Sekuler, A. B., Gaspar, C. M., Gold, J. M., & Bennett, P. J. (2004). Inversion leads to quantitative, not qualitative, changes in face processing. *Current Biology, 14*, 391–396. http://dx.doi.org/10.1016/j.cub.2004.02.028

Senju, A., & Johnson, M. H. (2009). The eye contact effect: Mechanisms and development. *Trends in Cognitive Sciences, 13*, 127–134. http://dx.doi.org/10.1016/j.tics.2008.11.009

Somppi, S., Törnqvist, H., Hänninen, L., Krause, C., & Vainio, O. (2012). Dogs do look at images: Eye tracking in canine cognition research. *Animal Cognition, 15*, 163–174. http://dx.doi.org/10.1007/s10071-011-0442-1

Somppi, S., Törnqvist, H., Hänninen, L., Krause, C. M., & Vainio, O. (2014). How dogs scan familiar and inverted faces: An eye movement study. *Animal Cognition, 17*, 793–803. http://dx.doi.org/10.1007/s10071-013-0713-0

Soproni, K., Miklósi, Á., Topál, J., & Csányi, V. (2001). Comprehension of human communicative signs in pet dogs (*Canis familiaris*). *Journal of Comparative Psychology, 115*, 122–126. http://dx.doi.org/10.1037/0735-7036.115.2.122

Téglás, E., Gergely, A., Kupán, K., Miklósi, Á., & Topál, J. (2012). Dogs' gaze following is tuned to human communicative signals. *Current Biology, 22*, 209–212. http://dx.doi.org/10.1016/j.cub.2011.12.018

Tomasello, M., Hare, B., Lehmann, H., & Call, J. (2007). Reliance on head versus eyes in the gaze following of great apes and human infants: The cooperative eye hypothesis. *Journal of Human Evolution, 52*, 314–320. http://dx.doi.org/10.1016/j.jhevol.2006.10.001

Tomonaga, M. (2007). Visual search for orientation of faces by a chimpanzee (*Pan troglodytes*): Face-specific upright superiority and the role of facial configural properties. *Primates, 48*, 1–12. http://dx.doi.org/10.1007/s10329-006-0011-4

Topál, J., Gácsi, M., Miklósi, Á., Virányi, Z., Kubinyi, E., & Csányi, V. (2005). Attachment to humans: A comparative study on hand-reared wolves and differently socialized dog puppies. *Animal Behaviour, 70*, 1367–1375. http://dx.doi.org/10.1016/j.anbehav.2005.03.025

Valentine, T. (1988). Upside-down faces: A review of the effect of inversion upon face recognition. *British Journal of Psychology, 79*, 471–491. http://dx.doi.org/10.1111/j.2044-8295.1988.tb02747.x

Vilà, C., Savolainen, P., Maldonado, J. E., Amorim, I. R., Rice, J. E., Honeycutt, R. L., . . . Wayne, R. K. (1997). Multiple and ancient origins of the domestic dog. *Science, 276*(5319), 1687–1689.

Williams, F. J., Mills, D. S., & Guo, K. (2011). Development of a head-mounted, eye-tracking system for dogs. *Journal of Neuroscience Methods, 194*, 259–265. http://dx.doi.org/10.1016/j.jneumeth.2010.10.022

Willis, J., & Todorov, A. (2006). First impressions: Making up your mind after a 100-ms exposure to a face. *Psychological Science, 17*, 592–598. http://dx.doi.org/10.1111/j.1467-9280.2006.01750.x

3

HUMAN–ANIMAL INTERACTION AND THE DEVELOPMENT OF EXECUTIVE FUNCTIONS

DAPHNE S. LING, MELISSA KELLY, AND ADELE DIAMOND

Children, indeed, people of all ages, are affected in many positive ways by interacting with pets and companion animals. Human–animal interaction (HAI) can be socially, emotionally, physically, and cognitively beneficial. Our particular speciality happens to be executive functions (EFs), those vitally important mental processes that form the basis for strategizing, paying attention, prioritizing competing tasks, maintaining self-control, and adjusting to ever-changing situations. In this chapter, we examine the potential effects of HAI on EFs, starting with a broad overview of what EFs are and factors that influence them. Although there is currently little, if any, research that specifically demonstrates the link between EFs and HAI, the potential benefits of HAI for EFs hold promise for enhancing EF functioning in both children and adults.

http://dx.doi.org/10.1037/14856-004
The Social Neuroscience of Human–Animal Interaction, L. S. Freund, S. McCune, L. Esposito, N. R. Gee, and P. McCardle (Editors)

EFs are top-down processes we recruit when it would be insufficient, detrimental, or impossible to go on autopilot or rely on habitual or instinctive reactions. EFs are also sometimes referred to as *executive control* or *cognitive control* (Espy, 2004; E. K. Miller & Cohen, 2001). The three core EFs (Diamond, 2013; Miyake et al., 2000) are *inhibitory control* (including self-control and attentional control), *working memory* (WM), and *cognitive flexibility* (also known as *mental flexibility* or *set shifting*). From those core EFs, higher order EFs (e.g., planning, reasoning, problem solving) are built (e.g., Collins & Koechlin, 2012; Lunt et al., 2012). Because they make it possible for us to think before we act, resist temptations, stay focused, mentally play with ideas, reason, and quickly adapt to changed circumstances, EFs are predictive of achievement, health, wealth, and quality of life throughout life, often more so than IQ or socioeconomic status (Moffitt, 2012; Moffitt et al., 2011).

Inhibition

Inhibitory control (also called *inhibition*) involves resisting a strong inclination to do one thing and instead doing what is most appropriate or needed (Diamond, 2013). It includes the self-control to resist temptations and not act impulsively (e.g., thinking before you speak or act so you don't do something you'd regret or put your foot in your mouth; waiting before rushing to judgment). It also includes the discipline to stay on task and complete what you started, resisting the temptation to quit because you're frustrated, bored, or more fun things are calling, and continuing to work even though the reward may be a long time in coming. It also includes screening out distractions so that you are able to concentrate, pay attention, and stay focused. Self-regulation overlaps largely, though not completely, with inhibitory control (Diamond, 2013).

Working Memory

WM refers to our ability to actively hold information in mind and work with that information—for example, reordering a set of items you are holding in mind, updating information, or doing mental arithmetic. WM is crucial for making sense of anything that unfolds over time, for that always requires holding in mind what happened earlier and relating that to what is happening now (e.g., keeping track of a conversation, relating what you are reading now to what you read earlier, understanding the relation between a later effect and an earlier cause).

Cognitive Flexibility or Switching

Cognitive flexibility refers to the ability to look at the same thing in different ways, from different perspectives, to think outside the box. If your way of solving a problem is not working, cognitive flexibility can help you conceive of the problem in a different way or approach it from a different angle. Cognitive flexibility builds on the other two core EFs and comes in later in development (Davidson, Amso, Anderson, & Diamond, 2006; Garon, Bryson, & Smith, 2008). Successful and fluid switching requires activating a new mind-set (loading it into WM) and deactivating (or inhibiting) the old mind-set. It is in this sense that cognitive flexibility requires and builds on inhibitory control and WM. Cognitive flexibility is also critical for adapting to change. You may be called on to flexibly change your plans when the unexpected happens, whether it is an unanticipated problem or obstacle or an unexpected opportunity or offer. Although the evidence for switching difficulty is most robust in early childhood and old age, even young adults in their prime find switching to be a challenge (Diamond & Kirkham, 2005; Koch, Gade, Schuch, & Philipp, 2010).

Researchers have agreed that the three constructs—inhibition, WM, and cognitive flexibility—are interrelated and interdependent. An improvement in either WM or inhibition is likely to lead to improvements in all three.

EVIDENCE ON THE IMPORTANCE OF EFs
FOR DIVERSE ASPECTS OF LIFE

Healthy EF development is one of the most critical developmental tasks. One of the first places that a young child is called on to exercise EFs is at school entry, when he or she is required to remember and follow instructions and inhibit the inclinations to not wait in line, take what he or she wants even if another child happens to have it at the moment, or speak whenever a thought pops into mind. EFs have been shown to be more important for school readiness than IQ or entry-level math or reading skills (e.g., Alloway et al., 2005; Blair, 2002).

EFs are important for success throughout the school years, from the earliest grades through university (often more so than IQ; Duckworth & Seligman, 2005; Gathercole, Pickering, Knight, & Stegmann, 2004; Nicholson, 2007). This is hardly surprising considering that skills valued in academic settings (e.g., discipline, conscientiousness, not always putting one's needs and desires ahead of everyone else's, following instructions, finishing assigned tasks) rely heavily on EFs (Diamond, 2014a).

The importance of strong EFs does not stop in childhood. There is abundant evidence that EFs are crucial for success later in life in the workplace (Bailey, 2007), marriage (Eakin et al., 2004), weight control (Crescioni et al., 2011), staying out of jail (Moffitt et al., 2011), and resisting substance abuse (H. V. Miller, Barnes, & Beaver, 2011). Adults with better EFs also report they are happier and have a better quality of life (Moffitt, 2012).

In a study of 1,000 children born in the same city and same year, those with worse inhibitory control (children who were less persistent, were more impulsive, and had poorer attention regulation) when they were young, as adolescents were more likely to smoke, have unplanned pregnancies and drop out of school; as adults 30 years later they were likely to earn less, have worse health (3 times more likely to be addicted to drugs), be a single parent (twice as likely), or commit a crime (4 times more likely) than those who had had better inhibitory control as children, controlling for IQ, gender, social class, home lives, and family circumstances growing up (Moffitt et al., 2011). That is consistent with other evidence that early EF gains can reduce the later incidence of school failure, substance abuse and addictions, aggression, crime, other antisocial or inappropriate behaviors, and early death (Nagin & Tremblay, 1999; Vitaro, Barker, Brendgen, & Tremblay, 2012).

INTERVENING TO IMPROVE EFs

Improving EFs early is important because early intervention is far more effective and far less costly than trying to correct problems once they have developed. Economists have estimated a 16% to 18% return on investment from such early intervention (Rolnick & Grunewalk, 2007; Sege, 2011). Being able to enhance EFs early in a child's life is critical because it affects the trajectory (the negative or positive feedback loop) on which a child gets launched. Improving EF skills early gets children started on a trajectory for success. Conversely, letting children's EFs remain poor gets them started on a negative trajectory that can be extremely difficult and expensive to reverse. Indeed, it is quite likely that "interventions that achieve even small improvements in [the inhibitory control] for individuals could shift the entire distribution of outcomes in a salutary direction and yield large improvements in health, wealth, and crime rate for a nation" (Moffitt et al., 2011, p. 2694).

Happily, it is absolutely clear that EFs can be improved, even in very young children (e.g., Blair & Raver, 2014; Diamond, Barnett, Thomas, & Munro, 2007; Röthlisberger, Neuenschwander, Cimeli, Michel, & Roebers, 2012). A surprisingly diverse array of activities each have at least one research study published in a peer-reviewed journal showing that they improve EFs, such as computerized training, martial arts, aerobic activities, yoga, meditation, noncomputerized games, and certain school curricula (for reviews, see Diamond & Lee, 2011; Diamond & Ling, in press). Regardless of the type of activity, all activities that improve EFs have certain things in common, discussed here briefly.

- *Transfer of EF training is narrow.* Although improvement of EFs transfers, the transfer is narrow (Diamond & Lee, 2011; Melby-Lervåg & Hulme, 2013; Park, Gutchess, Meade, & Stine-Morrow, 2007). People improve on the skills they train on, and those improvements transfer to other contexts where those same skills are needed, but people only improve on what they practice; improvement does not transfer to other unpracticed skills. For example, training on WM improves WM but not self-control, creativity, or flexibility. Thus, to see widespread benefits, diverse EF skills should be trained and practiced; then narrow transfer of each results in widespread gains across those skills. Real-world activities such as martial arts (e.g., Lakes & Hoyt, 2004) and certain school curricula (e.g., Blair & Raver, 2014; Diamond et al., 2007; Raver et al., 2011; Riggs, Greenberg, Kusché, & Pentz, 2006) train diverse EF skills and have shown more widespread benefits than targeted computerized training.
- *A lot of practice is needed.* When studying what makes an expert across an array of fields, Ericsson, Nandagopal, and Roring (2009) found that regardless of the field, it takes many, many hours of practice (they said 10,000 hours) to become expert at something. There is no substitute for putting in lots of hours. The same holds true for EFs: For EFs to improve, a lot of practice is needed (e.g., Bergman Nutley et al., 2011; Davis et al., 2011).
- *Progressively greater challenges to EFs are needed.* EFs need to be continually challenged, not just used, for improvements to continue to be seen. Ericsson et al. (2009) found that to get really good at anything not only takes many hours of practice but also practice of a certain kind: always pushing yourself to go just past the limits of your comfort zone (practice within what Vygotsky [1978] called the "zone of proximal development"). When one group is randomly assigned to EF training with difficulty continuously and incrementally increasing and another group to EF training where difficulty does not increase, researchers always find that those progressively challenged show more EF improvement than those who keep training at the same level (e.g., Holmes, Gathercole, & Dunning, 2009; Klingberg et al., 2005).
- *Benefits disappear when practice stops.* Studies have demonstrated that EF benefits can last months or even years, but they almost always grow smaller as the time since training increases (Ball et al.,

2002; Rueda, Checa, & Combita, 2012). It would be unrealistic to expect benefits to continue indefinitely once practice stops. For example, with repeated practice you might work up to being able to do 80 sit-ups at a time, but if you stop practicing, a year from now it is unlikely that you could do 80 sit-ups all at once.

■ *Those with the weakest EFs benefit most.* Children at risk start school with worse EFs than their more economically advantaged peers (Hackman & Farah, 2009; Sarsour et al., 2011) and fall progressively further behind each school year (O'Shaughnessy, Lane, Gresham, & Beebe-Frankenberger, 2003). Improving EFs early might nip that dynamic in the bud. In other words, early EF training is an excellent candidate for reducing inequality (because it should improve the EFs of the most needy children most), thus heading off gaps in achievement and health between more- and less-advantaged children.

BRAIN REGIONS THAT UNDERLIE EFs

EFs begin developing during the first year of life (Cuevas, Swingler, Bell, Marcovitch, & Calkins, 2012; Diamond, 1991) and continue developing for over 2 decades (Crone, Wendelken, Donohue, van Leijenhorst, & Bunge, 2006; Davidson et al., 2006; Luna, 2009). EFs depend on neural circuits that include prefrontal cortex as a prominent node. Prefrontal cortex undergoes maturational changes even during infancy (e.g., Bell & Wolfe, 2007; Koenderink & Uylings, 1995) but takes over 2 decades to fully mature (e.g., Gogtay et al., 2004; Tamnes et al., 2013).

Other brain regions that play prominent roles in EF neural networks are the anterior cingulate cortex (Barber, Ursu, Stenger, & Carter, 2001; Milham, Banich, Claus, & Cohen, 2003), the parietal cortex (Dodds, Morein-Zamir, & Robbins, 2011; Olesen, Westerberg, & Klingberg, 2004), and the striatum (Lewis, Dove, Robbins, Barker, & Owen, 2004; Robbins, 2007).

SENSITIVITY OF PREFRONTAL CORTEX AND EFs TO EMOTIONAL, SOCIAL, AND PHYSICAL ASPECTS OF LIFE

Prefrontal cortex is the newest and most vulnerable region of the brain. If you are sad or stressed, lonely or socially isolated, sleep deprived, or not physically fit, prefrontal cortex and EFs suffer first and most. Conversely, you show much better EFs when you feel emotionally and socially nourished and your body is healthy.

Stress and Its Effects on EFs

Our brains work better when we are not in a stressed emotional state, and that is particularly true for prefrontal cortex and EFs (Arnsten, Mazure, & Sinha, 2012). One reason stress impairs EFs first and most is that even mild stress markedly increases levels of the neurotransmitter, dopamine, in prefrontal cortex but not elsewhere in the brain (overwhelming prefrontal cortex with dopamine so that it cannot function properly; Vijayraghavan, Wang, Birnbaum, Williams, & Arnsten, 2007). Indeed, stress can make us look like we have an EF disorder, such as attention-deficit/hyperactivity disorder, when we do not. One month of stress in preparation for a major exam disrupts prefrontal cortex functioning connectivity and impairs EFs (Liston, McEwen, & Casey, 2009). You may have noticed that when you are stressed, you cannot think as clearly or exercise as good self-control.

Sadness and Its Effects on EFs

When we are sad, we have worse attentional control (Desseilles et al., 2009; von Hecker & Meiser, 2005); when we are happy, we have better attentional control (Gable & Harmon-Jones, 2008). We are more creative, more likely to think outside the box, when we are happier (Ashby, Isen, & Turken, 1999; Hirt, Devers, & McCrea, 2008; Isen, Daubman, & Nowicki, 1987). (It's not that happier people are more creative than sadder people, but that in general a given individual tends to be more creative when he or she is happier than when he or she is more miserable.)

Loneliness and Its Effects on EFs

When we are lonely, our EFs also suffer (e.g., Cacioppo & Patrick, 2008; Campbell et al., 2006). When we feel more socially supported and less isolated, we show better EFs (Cacioppo & Patrick, 2008; Tangney, Baumeister, & Boone, 2004). Social relationships nourish us in many ways and even affect gene expression (Szyf, McGowan, & Meaney, 2008) and the way our neural networks behave, grow, and adapt to change (Jablonka & Lamb, 2006). We are fundamentally social; children who are lonely or ostracized are usually less able to show the EFs of which they are capable.

Exercise and Its Effects on EFs

EFs tend to be poorer in sedentary individuals (Hillman, Erickson, & Kramer, 2008). Physical activity, especially if it requires the use of EFs (e.g., yoga, Manjunath & Telles, 2001; or taekwondo, Lakes & Hoyt, 2004) improves

EFs. Exercise also improves mood (Lane & Lovejoy, 2001; Williamson, Dewey, & Steinberg, 2001) and sleep (Foti, Eaton, Lowry, & McKnight-Ely, 2011; Loprinzi & Cardinal, 2011), which are two avenues by which it might improve EFs, in addition to when it directly taxes them.

THEORY PROPOSED BY DIAMOND

Diamond (2012, 2013, 2014b) proposed that the activities and programs that most successfully improve EFs are not only those that directly train and challenge EFs but also those that indirectly support EFs by helping to reduce things that disrupt them (e.g., stress) and/or by increasing things that aid EFs (e.g., social support; see Figure 3.1). What activities directly train and challenge EFs and indirectly support them by also addressing social, emotional, and physical needs? Some of the activities that do that best are the arts (e.g., music making, dance, theatre), physical activities (e.g., team sports, martial arts, youth circus), and caring for animals. All of these train and challenge multiple EFs; they also relieve stress and provide joy, companionship, and physical exercise. Therefore, we predict that they should be excellent

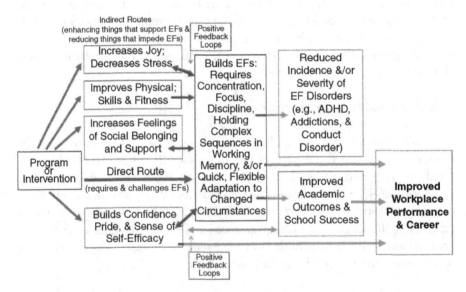

Figure 3.1. Illustration of a model put forward by Diamond (2012, 2013, 2014b) that postulates (a) that the activities that will most successfully improve EFs are those that require and directly challenge EFs but also indirectly support EFs by reducing things that impair them (e.g., stress) and enhancing things that support them (e.g., social belonging) and (b) that from that, better school and workplace performance and the reduced incidence and/or severity of EF disorders will follow.

candidates for improving EFs. Key is that an individual really enjoy the activity and really want to do it, so he or she will spend a lot of time at it, pushing himself or herself to improve. The best possible reason for anyone to want to keep at a difficult task is that he or she enjoys it.

CARING FOR AN ANIMAL REQUIRES EFs

Taking care of an animal provides many opportunities for the daily exercise of EFs. The responsibility of caring for a pet—remembering to walk, feed, groom, and keep the animal healthy—is an opportunity to exercise and train WM and inhibitory control. Continuous pet care over a long period (pets never grow up to be self-sufficient) taps inhibitory control and WM (focusing on a pet's needs even when you'd rather be doing something else or feel pressed for time, such as remembering when to give medications or to not place delicious human food too easily within reach of a pet, or forcing oneself to do things that are not so appealing, such as cleaning up poop). Training and taking care of an animal requires consistency.

Caring for an animal's needs provides opportunities for learning and exercising responsibility, even in very young children, and responsibilities can be incremented over time (gradually taxing EFs more and more). Young children can be taught, for example, to feed the family pet at specific times or to check that the pet has enough water. Children can also be involved in following the vet's instructions for giving medications and/or keeping the pet on a special diet. Repeatedly practicing responsible behaviors in caring for an animal can help make those behaviors more engrained and second nature. If you perceive yourself as a responsible pet owner, you might come to see yourself as a responsible person in general and act more responsibly in other realms. In short, within one's familiar home environment, caring for a pet can provide many occasions for EF skills to be exercised and nurtured.

HAI REDUCES STRESS

Many dog owners feel safer having a dog at home. There is evidence that in the face of anxiety-provoking stimuli, an animal's presence can help shift our attention from feeling helpless to feeling more hopeful and able to do something about the anxiety (Brickel, 1982; Shiloh, Sorek, & Terkel, 2003). The act of stroking an animal can relieve stress. Having companion animals present during anxiety-provoking procedures helps reduce children's distress (in pediatricians' offices, Hansen, Messinger, Baun, & Megel, 1999; Nagengast, Baun, Megel, & Leibowitz, 1997; during hospitalization, Tsai,

Friedmann, & Thomas, 2010; in their own homes/neighborhoods, Bryant, 1985; or general distress [e.g., bad day], McNicholas & Collis, 2001). Pets (especially dogs) also act to mediate stress levels in children with insecure attachment relationships (Beetz, Julius, Turner, & Kotrschal, 2012; Beetz et al., 2011). There are indications that just having an animal in the room reduces anxiety, allowing energy to be used more constructively for the tasks at hand. An increasing body of evidence shows that HAIs have a positive influence in the workplace. For example, dogs in the workplace have been found to reduce employee stress and increase team cooperation (Barker, Knisely, Barker, Cobb, & Schubert, 2012).

HAI IMPROVES MOOD AND GIVES JOY

Animals, especially dogs, express such exuberant joy on seeing their owner or someone familiar that it cannot help but make the recipient, and often observers, happy as well. Baby animals are so cute that they often bring smiles to our faces. To see the joy of animals when they are playing or outdoors running can also lift a person's spirits. An animal will sometimes explicitly try to lift someone's spirits by licking or nuzzling the person. HAI can help us see ourselves as better people ("If Boxer is so happy to see me, maybe I'm not so bad after all"), causing us to feel happier about ourselves. When we give an animal joy (e.g., by petting his coat, rubbing his belly, or giving him exercise), we get the great sense of pleasure that comes from making someone else happy, and it seems to take so little to cause an animal joy.

The presence of a dog in the classroom has been shown to help children exercise better EFs and perform better academically. That effect is credited to the dog reducing stress and improving the mood of the people in the classroom (teacher and students), allowing energy to be used more constructively for the tasks at hand. Preschoolers require fewer instructional prompts, are better able to keep their attention focused on their work, make fewer errors, and work more efficiently when there is a dog in the classroom (Gee, Church, & Altobelli, 2010; Gee, Crist, & Carr, 2010; Gee, Harris, & Johnson, 2007). The presence of an animal has also been shown to help children with cognitive impairments (Nathanson, 1989; Netting, Wilson, & New, 1987), pervasive developmental disorder (Martin & Farnum, 2002), and Down syndrome (Limond, Bradshaw, & Cormack, 1997) sustain focused attention longer.

Equine therapy (also known as hippotherapy) can provide children who have physical challenges (e.g., who cannot walk unaided or are normally confined to a wheelchair) with a sense of autonomy and accomplishment ("I can do this!") and provide them with a sense of pride as they sit tall, heads above those standing on the ground (All, Loving, & Crane, 1999; Granados & Agís, 2011).

HAI EASES LONELINESS AND PROVIDES COMPANIONSHIP

For centuries, animals have been welcomed into our homes in exchange for their love and companionship. Veevers (1985) suggested that a companion animal can substitute for, or complement, human interaction and can facilitate human interaction by acting as a bridge between people.

Pets help their human owners feel less lonely (Barker et al., 2012; Hunt, Hart, & Gomulkiewicz, 1992). This has been seen in people living alone (Zasloff & Kidd, 1994), children with disabilities (Mader, Hart, & Bergin, 1989) and without (Bryant, 1985; Covert, Whirren, Keith, & Nelson, 1985; Melson & Schwarz, 1994), rural youth (Black, 2012), and the elderly (Scheibeck, Pallauf, Stellwag, & Seeberger, 2011).

Dogs do not hold a grudge but give us unconditional positive regard; they love us no matter what. They also give us someone to love. They are always there for us. We can talk to them to hear ourselves think out loud, vent emotions, or try out different scenarios. If we want to go outside, they are always delighted to accompany us. Dogs do not require eye contact; those who find eye contact overwhelming or stressful often find interacting with dogs easier and more pleasant.

Having a pet can facilitate social interaction and act as a bridge to human interaction. It can be easier to start a conversation with "Oh, what a cute dog!" and someone too timid to speak directly to another person will often speak to the dog that person is walking. Walking one's dog also provides an opportunity to get out and meet others and socialize with one's neighbors (Friedmann & Thomas, 1985). It has consistently been found that companion animals help older individuals navigate the social circle, and that acts as a buffer for their psychological well-being (Hart, 2006; Raina, Waltner-Toews, Bonnett, Woodward, & Abernathy, 1999).

It is noteworthy that children are better able to empathize and appreciate individual differences when they are attached to a pet (e.g., Thompson & Gullone, 2003; Walsh, 2009). We are more likely to prefer spending time with people who show empathy and appreciate nuances in our communication or personal style. Children benefit from more than just "having a friend"; in their relationships with their pets, they can safely practice social skills that they can then transfer to human relationships.

HAI CAN IMPROVE PHYSICAL FITNESS

Individuals with companion animals report higher levels of physical activity and fitness than those without animal companions (Raina et al., 1999; Scheibeck et al., 2011; Serpell, 1991; Yabroff, Troiano, & Berrigan,

2008). Physical activity (at least walking one's dog) becomes part of one's daily routine as a dog owner. Even if you don't feel like it or the weather is rotten, your dog needs to go out. Pet owners are more likely to have consistent physical activity throughout the week. Thus, attending to the basic needs of their dogs seems to help owners lead healthier lifestyles. Similarly, if you own a horse, your horse needs regular exercise several times a week. From an EF lens, it appears that pets can help get their human owners up and exercising, and with repeated practice that routine can become second nature.

Riding a horse also has other benefits for the body. It can help improve head and trunk stability (Champagne & Dugas, 2010; Shurtleff, Standeven, & Engsberg, 2009), gait and balance (Kwon et al., 2011), gross motor function (reviewed in Snider, Korner-Bitensky, Kammann, Warner, & Saleh, 2007; Sterba, 2007), and muscle strength symmetry (Benda, McGibbon, & Grant, 2003).

CONCLUSION

It would appear that HAI can readily provide natural opportunities for training, using, and practicing EF skills, while reducing stress and providing joy, companionship and social connection, and opportunities for physical activity. Thus, we predict that it should be an excellent way to improve EFs. The multilayered positive aspects of HAI for practicing inhibition and stretching WM, stress reduction, social skills building, and physical fitness hold the potential to create an extremely effective intervention. Given how important EFs are for success in all of life's aspects, such interventions are well worth exploring. No study has yet looked at whether HAI, in fact, improves EFs. Research is needed that not only looks at whether this effect exists, but also that examines causal mechanisms. Humans and animals have led intimately intertwined lives for millennia. Given what we know about optimizing EFs, it makes sense that HAI provides a positive context for fostering the development of EF skills.

REFERENCES

All, A. C., Loving, G. L., & Crane, L. L. (1999). Animals, horseback riding, and implications for rehabilitation therapy. *Journal of Rehabilitation, 65*, 49–57.

Alloway, T. P., Gathercole, S. E., Adams, A. M., Willis, C., Eaglen, R., & Lamont, E. (2005). Working memory and phonological awareness as predictors of progress towards early learning goals at school entry. *British Journal of Developmental Psychology, 23*, 417–426. http://dx.doi.org/10.1348/026151005X26804

Arnsten, A., Mazure, C. M., & Sinha, R. (2012). This is your brain in meltdown. *Scientific American, 306,* 48–53. http://dx.doi.org/10.1038/scientificamerican 0412-48

Ashby, F. G., Isen, A. M., & Turken, A. U. (1999). A neuropsychological theory of positive affect and its influence on cognition. *Psychological Review, 106,* 529–550. http://dx.doi.org/10.1037/0033-295X.106.3.529

Bailey, C. E. (2007). Cognitive accuracy and intelligent executive function in the brain and in business. *Annals of the New York Academy of Sciences, 1118,* 122–141. http://dx.doi.org/10.1196/annals.1412.011

Ball, K., Berch, D. B., Helmers, K. F., Jobe, J. B., Leveck, M. D., Marsiske, M., . . . Willis, S. L.; for the Advanced Cognitive Training for Independent and Vital Elderly Study Group. (2002). Effects of cognitive training interventions with older adults: A randomized controlled trial. *JAMA, 288,* 2271–2281. http:// dx.doi.org/10.1001/jama.288.18.2271

Barber, A. D., Ursu, S., Stenger, V. A., & Carter, C. S. (2001, April). *Dorsolateral prefrontal and anterior cingulate cortex contributions to cognitive control.* Paper presented at the annual meeting of the Cognitive Neuroscience Society.

Barker, R. T., Knisely, J. S., Barker, S. B., Cobb, R. K., & Schubert, C. M. (2012). Preliminary investigation of employee's dog presence on stress and organizational perceptions. *International Journal of Workplace Health Management, 5,* 15–30. http://dx.doi.org/10.1108/17538351211215366

Beetz, A., Julius, H., Turner, D., & Kotrschal, K. (2012). Effects of social support by a dog on stress modulation in male children with insecure attachment. *Frontiers in Psychology, 3,* 352. http://dx.doi.org/10.3389/fpsyg.2012.00352

Beetz, A., Kotrschal, K., Turner, D., Hediger, K., Uvnäs-Moberg, K., & Julius, H. (2011). The effect of a real dog, toy dog and friendly person on insecurely attached children during a stressful task: An exploratory study. *Anthrozoös, 24,* 349–368. http://dx.doi.org/10.2752/175303711X13159027359746

Bell, M. A., & Wolfe, C. D. (2007). Changes in brain functioning from infancy to early childhood: Evidence from EEG power and coherence during working memory tasks. *Developmental Neuropsychology, 31,* 21–38. http://dx.doi.org/10.1080/ 87565640709336885

Benda, W., McGibbon, N. H., & Grant, K. L. (2003). Improvements in muscle symmetry in children with cerebral palsy after equine-assisted therapy (hippotherapy). *Journal of Alternative and Complementary Medicine, 9,* 817–825. http:// dx.doi.org/10.1089/107555303771952163

Bergman Nutley, S., Söderqvist, S., Bryde, S., Thorell, L. B., Humphreys, K., & Klingberg, T. (2011). Gains in fluid intelligence after training non-verbal reasoning in 4-year-old children: A controlled, randomized study. *Developmental Science, 14,* 591–601. http://dx.doi.org/10.1111/j.1467-7687.2010.01022.x

Black, K. (2012). The relationship between companion animals and loneliness among rural adolescents. *Journal of Pediatric Nursing, 27,* 103–112. http://dx.doi. org/10.1016/j.pedn.2010.11.009

Blair, C. (2002). School readiness: Integrating cognition and emotion in a neuro-biological conceptualization of children's functioning at school entry. *American Psychologist, 57,* 111–127. http://dx.doi.org/10.1037/0003-066X.57.2.111

Blair, C., & Raver, C. C. (2014). Closing the achievement gap through modification of neurocognitive and neuroendocrine function: Results from a cluster random-ized controlled trial of an innovative approach to the education of children in kindergarten. *PLoS ONE, 9,* e112393. http://dx.doi.org/10.1371/journal.pone.0112393

Brickel, C. M. (1982). Pet-facilitated psychotherapy: A theoretical explanation via attention shifts. *Psychological Reports, 50,* 71–74. http://dx.doi.org/10.2466/pr0.1982.50.1.71

Bryant, B. K. (1985). The neighborhood walk: Sources of support in middle child-hood. *Monographs of the Society for Research in Child Development, 50*(3), 34–44. http://dx.doi.org/10.2307/3333847

Cacioppo, J., & Patrick, W. (2008). *Loneliness: Human nature and the need for social connection.* New York, NY: Norton.

Campbell, W. K., Krusemark, E. A., Dyckman, K. A., Brunell, A. B., McDowell, J. E., Twenge, J. M., & Clementz, B. A. (2006). A magnetoencephalography investigation of neural correlates for social exclusion and self-control. *Social Neuroscience, 1,* 124–134. http://dx.doi.org/10.1080/17470910601035160

Champagne, D., & Dugas, C. (2010). Improving gross motor function and postural control with hippotherapy in children with Down syndrome: Case reports. *Physiotherapy Theory and Practice, 26,* 564–571. http://dx.doi.org/10.3109/09593981003623659

Collins, A., & Koechlin, E. (2012). Reasoning, learning, and creativity: Frontal lobe function and human decision-making. *PLoS Biology, 10*(3), e1001293. http://dx.doi.org/10.1371/journal.pbio.1001293

Covert, A. M., Whirren, A. P., Keith, J., & Nelson, C. (1985). Pets, early adolescents, and families. *Marriage & Family Review, 8,* 95–108. http://dx.doi.org/10.1300/J002v08n03_08

Crescioni, A. W., Ehrlinger, J., Alquist, J. L., Conlon, K. E., Baumeister, R. F., Schatschneider, C., & Dutton, G. R. (2011). High trait self-control predicts positive health behaviors and success in weight loss. *Journal of Health Psychology, 16,* 750–759. http://dx.doi.org/10.1177/1359105310390247

Crone, E. A., Wendelken, C., Donohue, S., van Leijenhorst, L., & Bunge, S. A. (2006). Neurocognitive development of the ability to manipulate information in working memory. *Proceedings of the National Academy of Sciences, USA, 103,* 9315–9320. http://dx.doi.org/10.1073/pnas.0510088103

Cuevas, K., Swingler, M. M., Bell, M. A., Marcovitch, S., & Calkins, S. D. (2012). Measures of frontal functioning and the emergence of inhibitory control pro-cesses at 10 months of age. *Developmental Cognitive Neuroscience, 2,* 235–243. http://dx.doi.org/10.1016/j.dcn.2012.01.002

Davidson, M. C., Amso, D., Anderson, L. C., & Diamond, A. (2006). Development of cognitive control and executive functions from 4 to 13 years: Evidence from manipulations of memory, inhibition, and task switching. *Neuropsychologia, 44,* 2037–2078. http://dx.doi.org/10.1016/j.neuropsychologia.2006.02.006

Davis, C. L., Tomporowski, P. D., McDowell, J. E., Austin, B. P., Miller, P. H., Yanasak, N. E., . . . Naglieri, J. A. (2011). Exercise improves executive function and achievement and alters brain activation in overweight children: A randomized, controlled trial. *Health Psychology, 30,* 91–98. http://dx.doi.org/10.1037/a0021766

Desseilles, M., Balteau, E., Sterpenich, V., Dang-Vu, T. T., Darsaud, A., Vandewalle, G., . . . Schwartz, S. (2009). Abnormal neural filtering of irrelevant visual information in depression. *The Journal of Neuroscience, 29,* 1395–1403. http://dx.doi.org/10.1523/JNEUROSCI.3341-08.2009

Diamond, A. (1991). Frontal lobe involvement in cognitive changes during the first year of life. In K. R. Gibson & A. C. Petersen (Eds.), *Brain maturation and cognitive development: Comparative and cross-cultural perspectives* (pp. 127–180). New York, NY: Aldine de Gruyter.

Diamond, A. (2012). Activities and programs that improve children's executive functions. *Current Directions in Psychological Science, 21,* 335–341. http://dx.doi.org/10.1177/0963721412453722

Diamond, A. (2013). Executive functions. *Annual Review of Psychology, 64,* 135–168. http://dx.doi.org/10.1146/annurev-psych-113011-143750

Diamond, A. (2014a). Understanding executive functions: What helps or hinders them and how executive functions and language development mutually support one another. *Perspectives on Language and Literacy, 40*(2), 7–11.

Diamond, A. (2014b). Want to optimize executive functions and academic outcomes? Simple, just nourish the human spirit. *Minnesota Symposia on Child Psychology, 37,* 205–232.

Diamond, A., Barnett, W. S., Thomas, J., & Munro, S. (2007). Preschool program improves cognitive control. *Science, 318,* 1387–1388. http://dx.doi.org/10.1126/science.1151148

Diamond, A., & Kirkham, N. (2005). Not quite as grown-up as we like to think: Parallels between cognition in childhood and adulthood. *Psychological Science, 16,* 291–297. http://dx.doi.org/10.1111/j.0956-7976.2005.01530.x

Diamond, A., & Lee, K. (2011). Interventions shown to aid executive function development in children 4 to 12 years of age. *Science, 333,* 959–964. http://dx.doi.org/10.1126/science.1204529

Diamond, A., & Ling, D. (in press). Fundamental questions surrounding efforts to improve executive functions (including working memory). In M. Bunting, J. Novick, M. Dougherty, & R. W. Engle (Eds.), *An integrative approach to cognitive and working memory training: Perspectives from psychology, neuroscience, and human development.* New York, NY: Oxford University Press.

Dodds, C. M., Morein-Zamir, S., & Robbins, T. W. (2011). Dissociating inhibition, attention, and response control in the frontoparietal network using functional magnetic resonance imaging. *Cerebral Cortex, 21,* 1155–1165. http://dx.doi.org/10.1093/cercor/bhq187

Duckworth, A. L., & Seligman, M. E. P. (2005). Self-discipline outdoes IQ in predicting academic performance of adolescents. *Psychological Science, 16,* 939–944. http://dx.doi.org/10.1111/j.1467-9280.2005.01641.x

Eakin, L., Minde, K., Hechtman, L., Ochs, E., Krane, E., Bouffard, R., . . . Looper, K. (2004). The marital and family functioning of adults with ADHD and their spouses. *Journal of Attention Disorders, 8,* 1–10. http://dx.doi.org/10.1177/108705470400800101

Ericsson, K. A., Nandagopal, K., & Roring, R. W. (2009). Toward a science of exceptional achievement: Attaining superior performance through deliberate practice. *Annals of the New York Academy of Sciences, 1172,* 199–217. http://dx.doi.org/10.1196/annals.1393.001

Espy, K. A. (2004). Using developmental, cognitive, and neuroscience approaches to understand executive control in young children. *Developmental Neuropsychology, 26,* 379–384. http://dx.doi.org/10.1207/s15326942dn2601_1

Foti, K. E., Eaton, D. K., Lowry, R., & McKnight-Ely, L. R. (2011). Sufficient sleep, physical activity, and sedentary behaviors. *American Journal of Preventive Medicine, 41,* 596–602. http://dx.doi.org/10.1016/j.amepre.2011.08.009

Friedmann, E., & Thomas, S. A. (1985). Health benefits of pets for families [Special issue: Pets and the family]. *Marriage & Family Review, 8*(3-4), 191–203. http://dx.doi.org/10.1300/J002v08n03_14

Gable, P. A., & Harmon-Jones, E. (2008). Approach-motivated positive affect reduces breadth of attention. *Psychological Science, 19,* 476–482. http://dx.doi.org/10.1111/j.1467-9280.2008.02112.x

Garon, N., Bryson, S. E., & Smith, I. M. (2008). Executive function in preschoolers: A review using an integrative framework. *Psychological Bulletin, 134,* 31–60. http://dx.doi.org/10.1037/0033-2909.134.1.31

Gathercole, S. E., Pickering, S. J., Knight, C., & Stegmann, Z. (2004). Working memory skills and educational attainment: Evidence from national curriculum assessments at 7 and 14 years of age. *Applied Cognitive Psychology, 18,* 1–16. http://dx.doi.org/10.1002/acp.934

Gee, N. R., Church, M. T., & Altobelli, C. L. (2010). Preschoolers make fewer errors on an object categorization task in the presence of a dog. *Anthrozoös, 23,* 223–230. http://dx.doi.org/10.2752/175303710X12750451258896

Gee, N. R., Crist, E. N., & Carr, D. N. (2010). Preschool children require fewer instructional prompts to perform a memory task in the presence of a dog. *Anthrozoös, 23,* 173–184. http://dx.doi.org/10.2752/175303710X12682332910051

Gee, N. R., Harris, S. L., & Johnson, K. (2007). The role of therapy dogs in speed and accuracy to complete motor skills tasks for preschool children. *Anthrozoös, 20,* 375–386. http://dx.doi.org/10.2752/089279307X245509

Gogtay, N., Giedd, J. N., Lusk, L., Hayashi, K. M., Greenstein, D., & Vaituzis, C., . . . Thompson, P. M. (2004). Dynamic mapping of human cortical development during childhood through early adulthood. *Proceedings of the National Academy of Sciences, USA, 101*, 8174–8179.

Granados, A. C., & Agís, I. F. (2011). Why children with special needs feel better with hippotherapy sessions: A conceptual review. *Journal of Alternative and Complementary Medicine, 17*, 191–197. http://dx.doi.org/10.1089/acm.2009.0229

Hackman, D. A., & Farah, M. J. (2009). Socioeconomic status and the developing brain. *Trends in Cognitive Sciences, 13*, 65–73. http://dx.doi.org/10.1016/j.tics.2008.11.003

Hansen, K. M., Messinger, C. J., Baun, M. M., & Megel, M. (1999). Companion animals alleviating distress in children. *Anthrozoös, 12*, 142–148. http://dx.doi.org/10.2752/089279399787000264

Hart, L. A. (2006). Community context and psychosocial benefits of animal companionship. In A. Fine (Ed.), *Handbook of animal-assisted therapy* (pp. 73–94). London, England: Academic Press. http://dx.doi.org/10.1016/B978-012369484-3/50006-2

Hillman, C. H., Erickson, K. I., & Kramer, A. F. (2008). Be smart, exercise your heart: Exercise effects on brain and cognition. *Nature Reviews Neuroscience, 9*, 58–65. http://dx.doi.org/10.1038/nrn2298

Hirt, E. R., Devers, E. E., & McCrea, S. M. (2008). I want to be creative: Exploring the role of hedonic contingency theory in the positive mood-cognitive flexibility link. *Journal of Personality and Social Psychology, 94*, 214–230. http://dx.doi.org/10.1037/0022-3514.94.2.94.2.214

Holmes, J., Gathercole, S. E., & Dunning, D. L. (2009). Adaptive training leads to sustained enhancement of poor working memory in children. *Developmental Science, 12*(4), F9–F15. http://dx.doi.org/10.1111/j.1467-7687.2009.00848.x

Hunt, S., Hart, L., & Gomulkiewicz, R. (1992). Role of small animals in social interactions between strangers. *The Journal of Social Psychology, 132*, 245–256. http://dx.doi.org/10.1080/00224545.1992.9922976

Isen, A. M., Daubman, K. A., & Nowicki, G. P. (1987). Positive affect facilitates creative problem solving. *Journal of Personality and Social Psychology, 52*, 1122–1131. http://dx.doi.org/10.1037/0022-3514.52.6.1122

Jablonka, E., & Lamb, M. J. (2006). *Evolution in four dimensions: Genetic, epigenetic, behavioral, and symbolic variation in the history of life.* Cambridge, MA: MIT Press.

Klingberg, T., Fernell, E., Olesen, P. J., Johnson, M., Gustafsson, P., Dahlström, K., . . . Westerberg, H. (2005). Computerized training of working memory in children with ADHD—a randomized, controlled trial. *Journal of the American Academy of Child & Adolescent Psychiatry, 44*, 177–186. http://dx.doi.org/10.1097/00004583-200502000-00010

Koch, I., Gade, M., Schuch, S., & Philipp, A. M. (2010). The role of inhibition in task switching: A review. *Psychonomic Bulletin & Review, 17*, 1–14. http://dx.doi.org/10.3758/PBR.17.1.1

Koenderink, M. J., & Uylings, H. B. (1995). Postnatal maturation of layer V pyramidal neurons in the human prefrontal cortex: A quantitative Golgi analysis.

Brain Research, 678, 233–243. http://dx.doi.org/10.1016/0006-8993(95)00206-6

Kwon, J.-Y., Chang, H. J., Lee, J. Y., Ha, Y., Lee, P. K., & Kim, Y.-H. (2011). Effects of hippotherapy on gait parameters in children with bilateral spastic cerebral palsy. *Archives of Physical Medicine and Rehabilitation*, 92, 774–779. http://dx.doi.org/10.1016/j.apmr.2010.11.031

Lakes, K. D., & Hoyt, W. T. (2004). Promoting self-regulation through school-based martial arts training. *Journal of Applied Developmental Psychology*, 25, 283–302. http://dx.doi.org/10.1016/j.appdev.2004.04.002

Lane, A. M., & Lovejoy, D. J. (2001). The effects of exercise on mood changes: The moderating effect of depressed mood. *The Journal of Sports Medicine and Physical Fitness*, 41, 539–545.

Lewis, S. J., Dove, A., Robbins, T. W., Barker, R. A., & Owen, A. M. (2004). Striatal contributions to working memory: A functional magnetic resonance imaging study in humans. *European Journal of Neuroscience*, 19, 755–760. http://dx.doi.org/10.1111/j.1460-9568.2004.03108.x

Limond, J. A., Bradshaw, J. W. S., & Cormack, M. K. F. (1997). Behavior of children with learning disabilities interacting with a therapy dog. *Anthrozoös*, 10, 84–89. http://dx.doi.org/10.2752/089279397787001139

Liston, C., McEwen, B. S., & Casey, B. J. (2009). Psychosocial stress reversibly disrupts prefrontal processing and attentional control. *Proceedings of the National Academy of Sciences, USA*, 106, 912–917. http://dx.doi.org/10.1073/pnas.0807041106

Loprinzi, P. D., & Cardinal, B. J. (2011). Association between objectively-measured physical activity and sleep, NHANES 2005–2006. *Mental Health and Physical Activity*, 4(2), 65–69. http://dx.doi.org/10.1016/j.mhpa.2011.08.001

Luna, B. (2009). Developmental changes in cognitive control through adolescence. *Advances in Child Development and Behavior*, 37, 233–278. http://dx.doi.org/10.1016/S0065-2407(09)03706-9

Lunt, L., Bramham, J., Morris, R. G., Bullock, P. R., Selway, R. P., Xenitidis, K., & David, A. S. (2012). Prefrontal cortex dysfunction and "Jumping to Conclusions": Bias or deficit? *Journal of Neuropsychology*, 6, 65–78. http://dx.doi.org/10.1111/j.1748-6653.2011.02005.x

Mader, B., Hart, L. A., & Bergin, B. (1989). Social acknowledgements for children with disabilities: Effects of service dogs. *Child Development*, 60, 1529–1534. http://dx.doi.org/10.2307/1130941

Manjunath, N. K., & Telles, S. (2001). Improved performance in the Tower of London test following yoga. *Indian Journal of Physiology and Pharmacology*, 45, 351–354.

Martin, F., & Farnum, J. (2002). Animal-assisted therapy for children with pervasive developmental disorders. *Western Journal of Nursing Research*, 24, 657–670. http://dx.doi.org/10.1177/019394502320555403

McNicholas, J., & Collis, G. M. (2001). Children's representations of pets in their social networks. *Child: Care, Health and Development, 27,* 279–294. http://dx.doi.org/10.1046/j.1365-2214.2001.00202.x

Melby-Lervåg, M., & Hulme, C. (2013). Is working memory training effective? A meta-analytic review. *Developmental Psychology, 49,* 270–291. http://dx.doi.org/10.1037/a0028228

Melson, G. F., & Schwarz, R. (1994). *Pets as social supports for families with young children.* Paper presented at the annual meeting of the Delta Society, New York, NY.

Milham, M. P., Banich, M. T., Claus, E. D., & Cohen, N. J. (2003). Practice-related effects demonstrate complementary roles of anterior cingulate and prefrontal cortices in attentional control. *NeuroImage, 18,* 483–493. http://dx.doi.org/10.1016/S1053-8119(02)00050-2

Miller, E. K., & Cohen, J. D. (2001). An integrative theory of prefrontal cortex function. *Annual Review of Neuroscience, 24,* 167–202. http://dx.doi.org/10.1146/annurev.neuro.24.1.167

Miller, H. V., Barnes, J. C., & Beaver, K. M. (2011). Self-control and health outcomes in a nationally representative sample. *American Journal of Health Behavior, 35,* 15–27. http://dx.doi.org/10.5993/AJHB.35.1.2

Miyake, A., Friedman, N. P., Emerson, M. J., Witzki, A. H., Howerter, A., & Wager, T. D. (2000). The unity and diversity of executive functions and their contributions to complex "Frontal Lobe" tasks: A latent variable analysis. *Cognitive Psychology, 41,* 49–100. http://dx.doi.org/10.1006/cogp.1999.0734

Moffitt, T. E. (2012, January). *Childhood self-control predicts adult health, wealth, and crime.* Paper presented at the Multi-Disciplinary Symposium Improving the Well-Being of Children and Youth, Copenhagen, Denmark.

Moffitt, T. E., Arseneault, L., Belsky, D., Dickson, N., Hancox, R. J., Harrington, H., . . . Caspi, A. (2011). A gradient of childhood self-control predicts health, wealth, and public safety. *Proceedings of the National Academy of Sciences, USA, 108,* 2693–2698. http://dx.doi.org/10.1073/pnas.1010076108

Nagengast, S. L., Baun, M. M., Megel, M., & Leibowitz, J. M. (1997). The effects of the presence of a companion animal on physiological arousal and behavioral distress in children during a physical examination. *Journal of Pediatric Nursing, 12,* 323–330. http://dx.doi.org/10.1016/S0882-5963(97)80058-9

Nagin, D., & Tremblay, R. E. (1999). Trajectories of boys' physical aggression, opposition, and hyperactivity on the path to physically violent and nonviolent juvenile delinquency. *Child Development, 70,* 1181–1196. http://dx.doi.org/10.1111/1467-8624.00086

Nathanson, D. E. (1989). Using Atlantic bottlenose dolphins to increase cognition of mentally retarded children. In P. Lovibond & P. Wilson (Eds.), *Clinical and abnormal psychology* (pp. 233–242). Amsterdam, The Netherlands: Elsevier.

Netting, F., Wilson, C., & New, J. (1987). The human-animal bond: Implications for practice. *Social Work, 32,* 60–64.

Nicholson, C. (2007, March 26). Beyond IQ: Youngsters who can focus on the task at hand do better in math. *Scientific American*. Retrieved from http://www.scientificamerican.com/article/beyond-iq-kids-who-can-focus-on-task-do-better-math/

Olesen, P. J., Westerberg, H., & Klingberg, T. (2004). Increased prefrontal and parietal activity after training of working memory. *Nature Neuroscience, 7*, 75–79. http://dx.doi.org/10.1038/nn1165

O'Shaughnessy, T., Lane, K. L., Gresham, F. M., & Beebe-Frankenberger, M. (2003). Children placed at risk for learning and behavioral difficulties: Implementing a school-wide system of early identification and prevention. *Remedial and Special Education, 24*, 27–35. http://dx.doi.org/10.1177/074193250302400103

Park, D. C., Gutchess, A. H., Meade, M. L., & Stine-Morrow, E. A. L. (2007). Improving cognitive function in older adults: Nontraditional approaches. *Journal of Gerontology, 62B*, 45–52. http://dx.doi.org/10.1093/geronb/62.special_issue_1.45

Raina, P., Waltner-Toews, D., Bonnett, B., Woodward, C., & Abernathy, T. (1999). Influence of companion animals on the physical and psychological health of older people: An analysis of a one-year longitudinal study. *Journal of the American Geriatrics Society, 47*, 323–329. http://dx.doi.org/10.1111/j.1532-5415.1999.tb02996.x

Raver, C. C., Jones, S. M., Li-Grining, C., Zhai, F., Bub, K., & Pressler, E. (2011). CSRP's impact on low-income preschoolers' preacademic skills: Self-regulation as a mediating mechanism. *Child Development, 82*, 362–378. http://dx.doi.org/10.1111/j.1467-8624.2010.01561.x

Riggs, N. R., Greenberg, M. T., Kusché, C. A., & Pentz, M. A. (2006). The mediational role of neurocognition in the behavioral outcomes of a social-emotional prevention program in elementary school students: Effects of the PATHS Curriculum. *Prevention Science, 7*, 91–102. http://dx.doi.org/10.1007/s11121-005-0022-1

Robbins, T. W. (2007). Shifting and stopping: Fronto-striatal substrates, neurochemical modulation and clinical implications. *Philosophical Transactions of the Royal Society of London. Series B: Biological Sciences, 362*, 917–932. http://dx.doi.org/10.1098/rstb.2007.2097

Rolnick, A. J., & Grunewalk, R. (2007). The economics of early childhood development as seen by two Federal Reserve Board economists. *Community Investments, 19*, 13–30.

Röthlisberger, M., Neuenschwander, R., Cimeli, P., Michel, E., & Roebers, C. M. (2012). Improving executive functions in 5- and 6-year-olds: Evaluation of a small group intervention in prekindergarten and kindergarten children. *Infant and Child Development, 21*, 411–429. http://dx.doi.org/10.1002/icd.752

Rueda, M. R., Checa, P., & Combita, L. M. (2012). Enhanced efficiency of the executive attention network after training in preschool children: Immediate changes and effects after two months. *Developmental Cognitive Neuroscience, 2*(Suppl. 1), S192–S204. http://dx.doi.org/10.1016/j.dcn.2011.09.004

Sarsour, K., Sheridan, M., Jutte, D., Nuru-Jeter, A., Hinshaw, S., & Boyce, W. T. (2011). Family socioeconomic status and child executive functions: The roles of language, home environment, and single parenthood. *Journal of the International Neuropsychological Society, 17*, 120–132. http://dx.doi.org/10.1017/S1355617710001335

Scheibeck, R., Pallauf, M., Stellwag, C., & Seeberger, S. (2011). Elderly people in many respects benefit from interaction with dogs. *European Journal of Medical Research, 16*, 557–563. http://dx.doi.org/10.1186/2047-783X-16-12-557

Sege, I. (2011). More evidence of long-lasting benefits. *Eye on Early Education: A Blog of Strategies for Children.* Retrieved from http://eyeonearlyeducation.org/2011/02/14/more-evidence-of-long-lasting-benefits/

Serpell, J. (1991). Beneficial effects of pet ownership on some aspects of human health and behaviour. *Journal of the Royal Society of Medicine, 84*, 717–720.

Shiloh, S., Sorek, G., & Terkel, J. (2003). Reduction of state-anxiety by petting animals in a controlled laboratory experiment. *Anxiety, Stress, & Coping, 16*, 387–395. http://dx.doi.org/10.1080/1061580031000091582

Shurtleff, T. L., Standeven, J. W., & Engsberg, J. R. (2009). Changes in dynamic trunk/head stability and functional reach after hippotherapy. *Archives of Physical Medicine and Rehabilitation, 90*, 1185–1195. http://dx.doi.org/10.1016/j.apmr.2009.01.026

Snider, L., Korner-Bitensky, N., Kammann, C., Warner, S., & Saleh, M. (2007). Horseback riding as therapy for children with cerebral palsy: Is there evidence of its effectiveness? *Physical & Occupational Therapy in Pediatrics, 27*(2), 5–23.

Sterba, J. A. (2007). Does horseback riding therapy or therapist-directed hippotherapy rehabilitate children with cerebral palsy? *Developmental Medicine and Child Neurology, 49*, 68–73. http://dx.doi.org/10.1017/S0012162207000175.x

Szyf, M., McGowan, P., & Meaney, M. J. (2008). The social environment and the epigenome. *Environmental and Molecular Mutagenesis, 49*, 46–60. http://dx.doi.org/10.1002/em.20357

Tamnes, C. K., Walhovd, K. B., Grydeland, H., Holland, D., Østby, Y., Dale, A. M., & Fjell, A. M. (2013). Longitudinal working memory development is related to structural maturation of frontal and parietal cortices. *Journal of Cognitive Neuroscience, 25*, 1611–1623. http://dx.doi.org/10.1162/jocn_a_00434

Tangney, J. P., Baumeister, R. F., & Boone, A. L. (2004). High self-control predicts good adjustment, less pathology, better grades, and interpersonal success. *Journal of Personality, 72*, 271–324. http://dx.doi.org/10.1111/j.0022-3506.2004.00263.x

Thompson, K. L., & Gullone, E. (2003). Promotion of empathy and prosocial behaviour in children through humane education. *Australian Psychologist, 38*, 175–182. http://dx.doi.org/10.1080/00050060310001707187

Tsai, C.-C., Friedmann, E., & Thomas, S. A. (2010). The effect of animal-assisted therapy on stress responses in hospitalized children. *Anthrozoös, 23*, 245–258. http://dx.doi.org/10.2752/175303710X12750451258977

Veevers, J. E. (1985). The social meaning of pets: Alternative roles for companion animals. *Marriage & Family Review, 8*(3-4), 11–30. http://dx.doi.org/10.1300/J002v08n03_03

Vijayraghavan, S., Wang, M., Birnbaum, S. G., Williams, G. V., & Arnsten, A. F. (2007). Inverted-U dopamine D1 receptor actions on prefrontal neurons engaged in working memory. *Nature Neuroscience, 10,* 376–384. http://dx.doi.org/10.1038/nn1846

Vitaro, F., Barker, E. D., Brendgen, M., & Tremblay, R. E. (2012). Pathways explaining the reduction of adult criminal behaviour by a randomized preventive intervention for disruptive kindergarten children. *Journal of Child Psychology and Psychiatry, 53,* 748–756. http://dx.doi.org/10.1111/j.1469-7610.2011.02517.x

von Hecker, U., & Meiser, T. (2005). Defocused attention in depressed mood: Evidence from source monitoring. *Emotion, 5,* 456–463. http://dx.doi.org/10.1037/1528-3542.5.4.456

Vygotsky, L. S. (1978). *Mind in society: The development of higher psychological processes.* Cambridge, MA: Harvard University Press.

Walsh, F. (2009). Human-animal bonds I: The relational significance of companion animals. *Family Process, 48,* 462–480. http://dx.doi.org/10.1111/j.1545-5300.2009.01296.x

Williamson, D., Dewey, A., & Steinberg, H. (2001). Mood change through physical exercise in nine- to ten-year-old children. *Perceptual and Motor Skills, 93,* 311–316. http://dx.doi.org/10.2466/pms.2001.93.1.311

Yabroff, K. R., Troiano, R. P., & Berrigan, D. (2008). Walking the dog: Is pet ownership associated with physical activity in California? *Journal of Physical Activity & Health, 5,* 216–228.

Zasloff, R. L., & Kidd, A. H. (1994). Loneliness and pet ownership among single women. *Psychological Reports, 75,* 747–752. http://dx.doi.org/10.2466/pr0.1994.75.2.747

INTEGRATIVE COMMENTARY I: DO COMPANION ANIMALS SUPPORT SOCIAL, EMOTIONAL, AND COGNITIVE DEVELOPMENT OF CHILDREN?

KURT KOTRSCHAL

Individual social connectedness, as well as success in school and society, is crucially supported by one's ability to control emotions and impulses, pursue goals (e.g., by keeping to agreements), and quickly and flexibly cope with changing conditions. This combination of skills and abilities is generally termed *executive functions* (EFs; Miyake et al., 2000), which also include forming and updating mental representations (Cunningham & Zelazo, 2007) and categories (Quinn, 2011); most of the functional-neurological connections of these psychological constructs run together in the prefrontal cortex (PFC; Miyake et al., 2000). EFs predict individual success in the private and public domains with good reliability (Diamond & Lee, 2011). The primary constructs of EF are inhibition (including self-control and self-regulation), working memory, and cognitive flexibility; higher order EFs are the ability to plan, reason, and problem solve (see Chapter 3, this volume).

http://dx.doi.org/10.1037/14856-005

The Social Neuroscience of Human–Animal Interaction, L. S. Freund, S. McCune, L. Esposito, N. R. Gee, and P. McCardle (Editors)

Partially overlapping with EF is the construct of social competence, with the ability to empathize at its core. In addition, a balanced emotionality has been noted to be a crucial variable for a long, happy, healthy life (Coan, 2011). EF, socioemotional and cognitive abilities, all of which lie at the interface of developmental psychology and neurobiology, are discussed in this integrative commentary of the chapters in this part (and in the volume generally), with an attempt to link these to the perspective of biophilia (Wilson, 1984) and attachment theory (Bowlby, 1969). Several of the chapters in this volume address the role of companion animals, the multitude of positive effects they can bring to their human partners, and the psychological and neurobiological influences such interactions have on child development (see Chapters 1 and 3 in this part, and Chapter 5).

People today increasingly experience societies as being out of balance, supported by frequent tales of epidemic depression, anxiety, and "burnout" (e.g., Embriaco et al., 2007). It remains unclear whether these reported impressions reflect increased incidence or increased awareness (Costello, Erkanli, & Angold, 2006). A key factor for developing and maintaining social and emotional balance and buffering against stress overload is the ability to bond to family members or those close (Larson, 2002). In fact, children seem increasingly to need assistance in recognizing and forming such bonds, at a time when psychological counseling for children is constrained by limited health care and education budgets (Newacheck & Kim, 2005). It is probably not too far-fetched to attribute these perceived societal changes to prevailing socioeconomic conditions and to electronic media competing for social relationship time. This is based on impression, not on data, because there are hardly any on this. Whatever the complex contexts, contingencies, and causalities may be, what can be done? Well-developed EFs, in addition to their important role in cognitive development, provide an optimal base for a balanced social and emotional life and can therefore help to counteract these societal pressures, thus making individuals more resilient. Therefore, any intervention that can support or enhance EF abilities, including animal assistance, may result in significant benefits for children's development. And, as Diamond stresses in her contribution in this part, the best interventions are those that are holistic in approach and that are not limited to the specific skills they are addressing. This definitely makes a case for animal-assisted pedagogy.

EVOLUTIONARY BACKGROUND: HUMAN INFANTS NEED TO DEVELOP THEIR FULL POTENTIAL

Contrary to what was proposed by the last century's school of behaviorism (e.g., Skinner, 1965), humans and other animals are not born with *tabula rasa* brains and minds, but with a relatively species-specific set of evolutionary "prefab" behaviors and perception schemes, including action patterns, sign stimuli (i.e., a stimulus that was selected to produce a particular behavioral response; Lorenz, 1943; Tinbergen, 1951), and behavior systems (Hinde, 1966), as well as mental templates, expectations, and learning potentials. Evolutionary relatedness between humans and animals is probably the main reason we share so many social and brain mechanisms with animals, for example, face perception (Chapter 2, this part).

One particularly important psychological trait in humans is the radically social construction of the human mind per se; babies may, in extreme cases, die or at least develop a severely constrained social phenotype for life if their earliest social needs are profoundly neglected (Ainsworth, Bell, & Stayton, 1971; Bowlby, 1969). Another, less universally agreed on trait, but an important one to consider in the context of a consideration of human–animal interaction (HAI), is "biophilia" (Wilson, 1984; Chapter 5, this volume). Finally, the development of representations is a third important trait: the way the brain perceives, depicts, and interprets the world is not objective in a scientific sense but is generated in an iterative process between the representations any brain already provides of the relevant features of this world (Paivio, 1990) and new experiences that cause updating of those representations (Cunningham & Zelazo, 2007). Indeed, it seems that children form (social) representation of animals by using the same brain mechanisms as used for representing humans (Chapter 1, this part). All these domains have important roles in individual social, emotional, cognitive, and bodily development, which may all benefit from interactions with animals.

Several environmental factors, including the presence of companion animals, may assist children in developing a healthy social connectedness and balanced emotionality. We discuss and integrate many important aspects of development with the information presented in chapters in this volume: Quinn (Chapter 1) presents findings on features of infants' mental representations of humans, animals, and objects. Guo (Chapter 2) stresses the common features of face perception in humans and animals. Ling, Kelly, and Diamond (Chapter 3) stress the neurobiological background on EFs and the developing brain, and consider how companion animals can provide emotional, social, physical, and cognitive benefits to developing children.

Human evolution includes mastering ecological and social challenges by means of a big, reflective, and philosophical brain, which in connection with complex human language is able to explain self in relation to the outside world not only in the here and now, but particularly over the past and into the future. This is true for different cultures as well as on the individual level. Therefore, scholars have concluded that features of

the human brain were selected to increase individual survival and repro-
ductive success in complex physical, ecological, and social environments.
Notwithstanding the debate as to how much of human cognitive ability
was directly selected for, and how much may be considered as emergent
properties of an increasingly complex central nervous system, these characteristics of the human brain guide interest and learning. This internal
faculty, termed *innate school-marm* by Konrad Lorenz (1965), or *instinct to
learn* by Peter Marler (1991), can be considered in perspective with the
concepts of representations and categories.

This faculty guiding the "interest system" (Panksepp, 2005), one of the
basic emotional systems shared with other animals, has evidently developed
over hundreds of thousands of years of attentive interaction between humans
and animals and their natural surroundings. Hence, it comes as little surprise
that the human philosophical brain seems to be biophilic by nature, that
is, instinctively and curiously interested in nature and animals (DeLoache,
Pickard, & LoBue, 2011; Wilson, 1984). This is impressively supported by
a number of findings from human ontogeny, which provide a valuable key
to human mental development over prehistoric times, because according to
Haeckel's Law, ontogeny at least coarsely tends to recap phylogeny (Haeckel,
1868/1883).

A NATURE-DEFICIT SYNDROME?

Human babies, for example, hold far longer attention spans to animals
compared with any other kind of objects (DeLoache et al., 2011). Also,
independent of culture, the first words children produce are usually animal
related, and tales and books for children are commonly dominated by animals; the younger the children, the greater their interest in animals (Wedl
& Kotrschal, 2009). Quinn (Chapter 1, this part) asserts that the human-like
qualities of animals are at least in part responsible for this attraction. Regardless
of the source of the attraction, contact with animals and nature can be considered an important component of child development. In fact, it has been proposed that children growing up in an environment of human artifacts without
contact to nature would develop a nature-deficit disorder (NDD; Louv, 2006).
Despite some evidence (Charles & Louv, 2009; Louv, 2006), NDD remains
a "soft" concept, notoriously hard to test, not officially acknowledged by the
Diagnostic and Statistical Manual of Mental Disorders (American Psychiatric
Association, 2013), yet not implausible. NDD includes attention-deficit disorder, anxiety, depression, and/or obesity, as well as weakly developed "folk
physics" and knowledge and judgments about nature, or weakness in making judgments in general. Whether or not NDD is a recognized syndrome,

a nearly identical list of associated symptoms might also be applicable for individuals with constrained EFs (Chapter 3, this part). In addition, getting children from a young age into regular, intense contact with nature has positive effects on mental development, including EFs. Similar positive effects are documented for animal-assisted pedagogy and therapy, which has been shown to be effective in improving most, if not all of these conditions (Beetz et al., 2011). Framed by the biophilia concept, these observations provide some plausibility, as does the information in the chapter by Ling and colleagues, that one of the important components for the optimal development of EFs is contact with nature and animals.

One reason for NDD or weakly developed EFs may be that, based on the human-specific evolutionary templates, mental representations develop in greater richness, diversity, and connectedness (qualitatively or quantitatively) when a child has certain age-specific experiences. It seems that over evolutionary history and throughout an individual child's development, the ability to relate oneself to the relevant features of the world develops in an iterative process: as they are perceived, additional features are integrated with preexisting mental representations. The brain mechanisms behind this process of dynamically updating mental representations are increasingly well-known (Cunningham & Zelazo, 2007); the representations of the items and contexts relevant for species and individuals are updated throughout life. Conversely, the new perceptions of, and experiences with, the environment are made (i.e., interpreted) on the basis of this filter mechanism, which amalgamates evolutionary templates with existing mental representations. The evident function of these mental-filter systems, turning the real world into a system of cognitive concepts, is guiding judgments of relevance and meaning. The evident function of this updating and guiding of judgments is to reconcile the individual self with its environment in an efficient way, which not only allows survival, but in evolutionary terms, enables the species to reproduce successfully.

Throughout a lifetime of interactions between sensory input and mental representations, individual brains are continuously modulated in how they interpret the world. The PFC plays a key role in this process (Cunningham & Zelazo, 2007). Therefore, the common belief that the most readily observable features of an object per se (shape, texture, motion) will determine its relevance, and subjective interpretation is an oversimplification. Experience is formed in an iterative process between stimulus construals (e.g., evolutionary templates) and interpretative processes affected by already existing representations (memory); in humans and other species with complex, cognition-based sociality (e.g., wolves, ravens), these representations will be heavily affected by life history and by social and cultural embedding.

Bonding between dependent offspring and mothers ensures caregiving (Curley & Keverne, 2005; Eibl-Eibesfeldt, 1970; Hinde, 1998) and protects the offspring from harm (Solomon & George, 2011). The central players in this process are the functional and responsive oxytocin system of the mother and the child's dynamic development in interaction with the primary care-giver. On the basis of this bond, human infants form an attachment relation-ship with the caregiver (Ainsworth et al., 1971; Bowlby, 1969). Depending on the mutual responsiveness of mother and infant, the primary social rela-tionship representation will vary between secure attachment (Bowlby, 1969) and disorganized attachment (Chapter 5, this volume; Solomon & George, 2011), shaping and affecting the child's expectations and social interactions throughout the life span, both physiologically and cognitively.

Secure attachment provides the individual with positive expectations of future social partners and the ability to engage in trustful and support-ive relationships later in life. Along with such secure attachment come well-calibrated physiological stress systems and a reactive oxytocin "calming" system (Chapter 4, this volume; Julius, Beetz, Kotrschal, Turner, & Uvnäs-Moberg, 2013; Uvnäs-Moberg, 2003); all are important foundations for the development of a balanced emotionality, temperament, and personality (Koolhaas et al., 2011), including well-developed EFs (Miyake et al., 2000), and ultimately for a long, successful, happy and healthy life (Coan, 2011; Diamond & Lee, 2011; Chapters 3 and 6, this volume).

Optimal social development prepares a child for, but is no substitute for, exploration and interaction with the objects and materials of nature and with animals as a base for developing folk physics and a sense of self in balanced perspective with the environment. It seems as if growing up with animals would benefit a child's physical, emotional, and cognitive development (i.e., supporting the development of EF) even in children with secure attachment (Beetz et al., 2011; Hergovich, Monshi, Semmler, & Zieglmayer, 2002; Julius et al., 2013; Kotrschal & Ortbauer, 2003; Chapter 5, this volume). Even securely attached children may show suboptimal social and cognitive devel-opment in comparison with their potential, if deprived of regular early con-tact with nature and animals; this may particularly affect creativity and EFs.

If the primary caregiver(s) acts unreliably, unsupportively, or inconsis-tently, children may develop insecure attachment, avoiding the caregiver in case of distress in favor of seeking alternative comfort, for example, in object play or by clinging insecurely to the caregiver. If things go severely wrong, as in cases of neglect, loss of caregiver, abuse, or other trauma, or if the child is severely unresponsive, a disorganized attachment representation may develop (Chapter 5, this volume). Maladaptive behaviors often result, impairing the individual's ability to receive and offer emotional support. Attachment disorganization affects approximately 15% of children (data from Germany; Julius et al., 2013) and generally is associated with suboptimally adjusted physiological stress and calming systems, impulsive or withdrawn temperament, and personality and emotional imbalance. The prevalence of learning impairment is also high, and EFs are negatively affected; individuals with impaired attachment tend to show constrained impulse control, insufficient behavioral inhibition, low social interest, impaired episodic memory, and some lack of consistency in pursuing their goals. These are skills needed for success at school and within society, and EF is indeed a better predictor of individual success at school and in life than IQ (Chapter 3, this volume).

However, humans are remarkably resilient; even individuals with disorganized attachment may later develop secure attachment via secondary social bonds or engage in compensatory strategies for activating their reward systems. It is particularly striking that animal assistance can positively affect individuals with attachment disorders. As noted earlier, EFs are core functions of the PFC (Cunningham & Zelazo, 2007; Diamond & Lee, 2011; Koechlin & Hyafil, 2007). In fact, the PFC, or its analogous equivalent in birds, is the "social control brain area" (Güntürkün, 2005). In mammals and particularly in humans, this youngest and most modern part of the neocortex is radically dependent on the social context for proper development, as the contingency between early caregiving, attachment, and EF development indicates. The factors benefitting optimal EF development are also supporting PFC development and functions, and these factors may include contact with nature and animals (Diamond & Lee, 2011).

ATTACHMENT-RELATED, ANIMAL–ASSISTED INTERVENTION

Animals may be a valuable asset not only in normal child development but particularly in the context of pedagogic and therapeutic interventions when things have gone wrong, as noted in Chapter 3, this volume. Others have also demonstrated that children with suboptimal attachment benefit from interactions with friendly animals but not with toy versions of those animals (Beetz et al., 2011), and that these results include changes in the stress hormone cortisol. This could be interpreted as showing that interaction with a friendly dog could, via a stress-dampening effect, trigger an upregulation of oxytocin (Julius et al., 2013; Uvnäs-Moberg, 2003). However, it may also be that children with disorganized attachment will re-enact their tendency to control others toward a pet, for example, a dog. Therefore, it remains to be investigated whether and to what extent suboptimal attachment and EF deficits, which are also here argued to be aspects of the NDD, may be stabilized or reversed by pedagogic or therapeutic contact with nature and animals. But it can be predicted that animal-assisted interventions are more likely to be successful the earlier in life they occur (Phoenix, Goy, Gerall, & Young, 1959).

AN INTEGRATED SYSTEM OF BASIC HUMAN REPRESENTATIONS?

The internal working model (Bowlby, 1969) of socioemotional development may be considered as the basic human social representation. It shows some stability over life, probably because it is formed very early and is based on an evolutionary template; virtually every newborn human has such an "instinct" to learn (Marler, 1991) about the quality of its primary caregiver.

On the basis of this and other basic human propensities, such as language learning, we might also assume that human biophilia comes with basic templates present already at or shortly after birth. This could be inferred from the strong interest of human babies in animals (DeLoache et al., 2011), decreasing and diversifying over development (Wedl & Kotrschal, 2009). Quinn also discusses human-typical categorization of the world along certain rules (2011; Chapter 1 in this volume) and provides insight into how such categorical representations of the real world are formed. Quinn argues, for example, from various evidence sources, including both behavioral and electroencephalographic data, that the contrast between how infants respond to humans versus nonhuman animals exemplifies an expert–novice difference in the early development of category representations. That is, infants may respond to animals similarly to the way they respond to humans, using a "like-people" foundation for a common-sense biology, implying that we form attachments to animals very early in life because we perceive them to be like ourselves. When presented with pictorial representations, however, Quinn notes that infants form categories based on online learning and gradually modify these categories iteratively as they gain additional experience. I would therefore argue that as babies seemingly distinguish between animals as potentially alive (even when shown via a picture) and toy animal models, one of the basic, template-based human (or even general vertebrate) representations should cover aliveness; aliveness, then, will be interpreted by individual subjects based on major interacting components, perception mechanisms, and evolutionary templates on the one hand and mental representations on the other.

The theory of biophilia-related representations serves as a reasonable model of the complex nature of the relationship between representations and categories. Once the general animalophilia of human babies has individualized and diversified via experience, individual experiences are used to interpret the behavior of subjects or even of inanimate objects. It would appear, thus, that the updating of the representations of the "real thing" is only possible on the basis of the already existing representation, the self, and our very subjective experience with it being part of these representations (Cunningham & Zelazo, 2007). Therefore, humans anthropomorphize and mentalize not only other sentient beings but also objects. *Anthropomorphism* is defined as the attribution of human characteristics or behavior to virtually all other entities in the environment (e.g., Guthrie, 1997) they relate to, be it pets, computers, or even gods. Whether and to what degree this happens with a particular animal or object will depend mainly on the mental representations thereof, which are built on personal experience. The evidence points at the involvement of both automatic and reflective processes in anthropomorphizing, as well as domain-specific and domain-general mechanisms that are recruited depending on the type of information available to the observer (Urquiza-Haas & Kotrschal, in press). Whether we accept the assumption of innate biophilia, the human representation system is profoundly socially intertwined, and early experience with animals influences how this representation is built.

What subject/object features will be most relevant may be the key question behind the finding that a real dog, but not a toy dog, was used by the children as a social support (Beetz et al., 2011; Julius et al., 2013). Contemporary approaches in behavioral biology toward defining the observable traits of "aliveness" depend on agency, such as the ability of a mouse to start and stop, in contrast to the passivity of a falling leaf (reviewed in Urquiza-Haas & Kotrschal, in press). It seems now widely accepted that visual perception features a kind of "animacy detector" specialized in discriminating living subjects from nonliving objects (Troje & Westhoff, 2006). On a theoretical base, one could assume that evolutionary encodings and templates inbuilt in perception mechanisms will be particularly affected by stimulus qualities of subjects/objects that were relevant over long periods of evolutionary history. Sign stimuli (Lorenz, 1943; Tinbergen, 1951) would directly reach subconscious levels of brain, potentially even bypassing mental representations (or penetrating their deep layers) and would, thereby, be most potent to affect behavioral responses and, secondarily, mental representations. A fair example of such simple sign stimuli would be *Kindchenschema* (Lorenz, 1943), otherwise known as pedomorphosis or neoteny (i.e., that child-like features, such as a round head, large eyes, and short bodily appendages are generally considered "cute" by most, also triggering caregiving behavior).

Another level of potential sign stimuli would be the way behavior is generally organized in fundamental motor units, the so-called action patterns (synonymously termed *innate behaviors*; Tinbergen, 1951); these are species specific and relatively stereotyped in shape but variable in intensity. Action patterns are typically and specifically linked with the motivational/affective background, which explains why they are expressed in specific contexts and intensities and in succession, in a nonrandom order. These rules are universal throughout the animal kingdom. It may therefore be assumed that a particular sign stimulus such as a perception mechanism has evolved. It was in fact shown that human mirror neurons respond to robots as well as to living models (Oberman, McCleery, Ramachandran, & Pineda, 2007). Whether, and to what degree, this opto-motor reflex network will be affected by mental representations is not known.

Working with the representations of aliveness and nature in children is a safe guess in creating a favorable environment for the optimal physical, emotional, and cognitive development of children. Over long periods of human evolution, nature and animals were firmly integrated with human social and spiritual life (Broom, 2003; Serpell, 1986) and therefore provided the environment that, we believe, the brain of a developing child will expect, based on innate templates. It would thus follow that certain aspects of nature and animals need to be considered to avoid NDD (Louv, 2006). Interestingly, Diamond and Lee (2011) did not highlight nature and animals as important agents in child development, but rather mentioned computerized training, noncomputerized games, aerobics, martial arts, yoga, mindfulness, and school curricula as important. In their contribution in this part, Ling, Kelly, and Diamond now also extensively include animals.

COMPANION ANIMALS SUPPORT CHILDREN IN DEVELOPING THEIR EXECUTIVE FUNCTIONS: REAL OR WISHFUL THINKING?

Engaging in this discussion and bringing companion animals into the picture has been an exciting exercise because of the many threads that can be integrated into a reasonably coherent picture of how early social environment affects sociability, stress coping, and the formation of representations, that is, what the optimal conditions are for the maturation of the PFC and therefore of EFs. This is possible today because of the amazing progress made in the last 2 decades, particularly in neurobiology and behavioral endocrinology. This makes it possible to sensibly integrate long-standing concepts of empirical pedagogics, such as those offered by Montessori, encouraging and allowing children to concentrate on their tasks in a calm and socially pleasant atmosphere. It seems clear that these are the conditions required for optimal EF development. We presently see an unprecedented convergence of results, of mechanisms and functions from areas as diverse as pedagogy, neurobiology, evolutionary biology, behavioral endocrinology, and psychology, and as a consequence, the breakdown of the borders between disciplines. Clearly, bottom-up interdisciplinarity emerges here.

But there is also a big caveat: The concise grand picture emerging from bringing together the threads from different disciplines may be treacherous. Different results may fit together well, based on phenomenology and scientific logic, but they have yet to be tied together by an adequate network of experiments that would allow us to definitively interpret relationships between pedagogy, neurobiology, psychological attachment theory, stress physiology, etc., as causal. What we have at the moment is an emerging grand synthetic theory on EF development (Julius et al., 2013) and how companion animals might fit in. On the basis of the work laid out in this volume, and in particular in the chapters in this part, it appears that the likelihood is substantial that we are discussing a causality and contingency network of theories, but more conclusive science still needs to be done. Our socialization with science suggests that in doing this science, we will gain lots of interesting and practically relevant twists and insights pertaining to the grand theory itself. The thrill is not over yet; it has just begun.

REFERENCES

Ainsworth, M. D. S., Bell, S. M., & Stayton, D. J. (1971). Individual differences in strange-situation behavior of one-year-olds. In H. R. Schaffer (Ed.), *The origins of human social relations* (pp. 17–52). New York, NY: Academic Press.

American Psychiatric Association. (2013). *Diagnostic and statistical manual of mental disorders* (5th ed.). Washington, DC: Author.

Beetz, A., Kotrschal, K., Turner, D. C., Hediger, K., Uvnäs-Moberg, K., & Julius, H. (2011). The effect of a real dog, toy dog and friendly person on insecurely attached children during a stressful task: An exploratory study. *Anthrozoös, 24,* 349–368. http://dx.doi.org/10.2752/175303711X13159027359746

Bowlby, J. (1969). *Attachment and loss: Vol. 1. Attachment.* New York, NY: Basic Books.

Broom, D. W. (2003). *The evolution of morality and religion.* Cambridge, England: Cambridge University Press. http://dx.doi.org/10.1017/CBO9780511610226

Charles, C., & Louv, R. (2009). *Children's nature deficit: What we know—and don't know.* Children & Nature Network. Retrieved from http://www.childrenand nature.org/wp-content/uploads/2015/04/CNNEvidenceoftheDeficit.pdf

Coan, J. A. (2011). The social regulation of emotion. In J. Decety & J. T. Cacioppo (Eds.), *Oxford handbook of social neuroscience* (pp. 614–623). New York, NY: Oxford University Press. http://dx.doi.org/10.1093/oxfordhb/9780195342161.013.0041

Costello, E. J., Erkanli, A., & Angold, A. (2006). Is there an epidemic of child or adolescent depression? *Journal of Child Psychology and Psychiatry, 47,* 1263–1271.

Cunningham, W. A., & Zelazo, P. D. (2007). Attitudes and evaluations: A social cognitive neuroscience perspective. *Trends in Cognitive Sciences, 11,* 97–104. http://dx.doi.org/10.1016/j.tics.2006.12.005

Curley, J. P., & Keverne, E. B. (2005). Genes, brains and mammalian social bonds. *Trends in Ecology & Evolution, 20,* 561–567. http://dx.doi.org/10.1016/j.tree.2005.05.018

DeLoache, J. S., Pickard, M. B., & LoBue, V. (2011). How very young children think about animals. In P. McCardle, S. McCune, J. A. Griffin, & V. Maholmes. (Eds.), *How animals affect us: Examining the influences of human–animal interaction on child development and human health* (pp. 85–99). Washington, DC: American Psychological Association. http://dx.doi.org/10.1037/12301-004

Diamond, A., & Lee, K. (2011). Interventions shown to aid executive function development in children 4 to 12 years old. *Science, 333,* 959–964. http://dx.doi.org/10.1126/science.1204529

Eibl-Eibesfeldt, I. (1970). *Liebe und Hafl. Zur Naturgeschichte elementarer Verhaltensweisen.* München, Germany: Piper.

Embriaco, N., Azoulay, E., Barrau, K., Kentish, N., Pochard, F., Loundou, A., & Papazian, L. (2007). High level of burnout in intensivists: Prevalence and associated factors. *American Journal of Respiratory and Critical Care Medicine, 175,* 686–692. http://dx.doi.org/10.1164/rccm.200608-1184OC

Güntürkün, O. (2005). The avian "prefrontal cortex" and cognition. *Current Opinion in Neurobiology, 15,* 686–693. http://dx.doi.org/10.1016/j.conb.2005.10.003

Guthrie, S. E. (1997). Anthropomorphism: A definition and a theory. In R. W. Mitchell, N. S. Thompson, & H. L. Miles (Eds.), *Anthropomorphism, anecdotes, and animals* (pp. 50–58). Albany: State University of New York Press.

Haeckel, E. (1883). *The history of creation* (3rd ed., Vol. 1; E. Ray Lankester, Trans.). London, England: Kegan Paul, Trench & Co. (Original work published 1868)

Hergovich, A., Monshi, B., Semmler, G., & Zieglmayer, V. (2002). The effects of the presence of a dog in the classroom. *Anthrozoös, 15,* 37–50. http://dx.doi.org/10.2752/089279302786992775

Hinde, R. A. (1966). *Animal behavior: A synthesis of ethology and comparative psychology.* New York, NY: McGraw-Hill.

Hinde, R. A. (1998). Mother-infant separation and the nature of inter-individual relationships: Experiments with rhesus monkeys. In J. Bolhuis & J. A. Hogan (Eds.), *The development of animal behaviour: A reader* (pp. 283–299). Oxford, England: Blackwell.

Julius, H., Beetz, A., Kotrschal, K., Turner, D., & Uvnäs-Moberg, K. (2013). *Attachment to pets.* New York, NY: Hogrefe.

Koechlin, E., & Hyafil, A. (2007). Anterior prefrontal function and the limits of human decision-making. *Science, 318,* 594–598. http://dx.doi.org/10.1126/science.1142995

Koolhaas, J. M., Bartolomucci, A., Buwalda, B., de Boer, S. F., Flügge, G., Korte, S. M., . . . Fuchs, E. (2011). Stress revisited: A critical evaluation of the stress concept. *Neuroscience and Biobehavioral Reviews, 35,* 1291–1301. http://dx.doi.org/10.1016/j.neubiorev.2011.02.003

Kotrschal, K., & Ortbauer, B. (2003). Behavioral effects of the presence of a dog in a classroom. *Anthrozoös, 16*, 147–159. http://dx.doi.org/10.2752/089279303786992170

Larson, R. W. (2002). Globalization, societal change, and new technologies: What they mean for the future of adolescence. *Journal of Research on Adolescence, 12*, 1–30. http://dx.doi.org/10.1111/1532-7795.00023

Lorenz, K. (1943). Die angeborenen Formen möglicher Erfahrung. *Zeitschrift für Tierpsychologie, 5*, 235–409. http://dx.doi.org/10.1111/j.1439-0310.1943.tb00655.x

Lorenz, K. (1965). *Evolution and modification of behavior.* Chicago, IL: University of Chicago Press.

Louv, R. (2006). Last child in the woods: Saving our children from nature-deficit disorder. *Children, Youth and Environments, 16*, 1546–2250.

Marler, P. (1991). Instinct to learn. In S. Carey & R. Gelman (Eds.), *The epigenesis of mind* (pp. 37–48). Hillsdale, NJ: Erlbaum.

Miyake, A., Friedman, N. P., Emerson, M. J., Witzki, A. H., Howerter, A., & Wager, T. D. (2000). The unity and diversity of executive functions and their contributions to complex "frontal lobe" tasks: A latent variable analysis. *Cognitive Psychology, 41*, 49–100. http://dx.doi.org/10.1006/cogp.1999.0734

Newacheck, P. W., & Kim, S. E. (2005). A national profile of health care utilization and expenditures for children with special health care needs. *Archives of Pediatrics & Adolescent Medicine, 159*, 10–17. http://dx.doi.org/10.1001/archpedi.159.1.10

Oberman, L. M., McCleery, J. P., Ramachandran, V. S., & Pineda, J. A. (2007). EEG evidence for mirror neuron activity during the observation of human and robot actions: Toward an analysis of the human qualities of interactive robots. *Neurocomputing, 70*, 2194–2203. http://dx.doi.org/10.1016/j.neucom.2006.02.024

Paivio, A. (1990). *Mental representations: A dual coding approach.* Oxford, England: Oxford University Press. http://dx.doi.org/10.1093/acprof:oso/9780195066661.001.0001

Panksepp, J. (2005). Affective consciousness: Core emotional feelings in animals and humans. *Consciousness and Cognition, 14*, 30–80. http://dx.doi.org/10.1016/j.concog.2004.10.004

Phoenix, C. H., Goy, R. W., Gerall, A. A., & Young, W. C. (1959). Organizing action of prenatally administered testosterone propionate on the tissues mediating mating behavior in the female guinea pig. *Endocrinology, 65*, 369–382. http://dx.doi.org/10.1210/endo-65-3-369

Quinn, P. C. (2011). Born to categorize. In U. Goswami (Ed.), *The Wiley-Blackwell handbook of childhood cognitive development* (pp. 129–152). Oxford, England: Blackwell.

Serpell, J. A. (1986). *In the company of animals.* Oxford, England: Blackwell.

Skinner, B. F. (1965). *Science and human behavior.* New York, NY: Free Press.

Solomon, J., & George, C. (2011). *Disorganization of attachment and caregiving.* New York, NY: Guilford Press.

Tinbergen, N. (1951). *The study of instinct*. Oxford, England: Clarendon Press/Oxford University.

Troje, N. F., & Westhoff, C. (2006). The inversion effect in biological motion perception: Evidence for a "life detector"? *Current Biology, 16*, 821–824. http://dx.doi.org/10.1016/j.cub.2006.03.022

Urquiza-Haas, E. G., & Kotrschal, K. (in press). The mind behind anthropomorphic thinking: Attribution of mental states to other species. *Animal Behaviour*.

Uvnäs-Moberg, K. (2003). *The oxytocin factor. Tapping the hormone of calm, love, and healing*. Cambridge, MA: Da Capo Press.

Wedl, M., & Kotrschal, K. (2009). Social and individual components of animal contact with animals in preschool children. *Anthrozoös, 22*, 383–396. http://dx.doi.org/10.2752/089279309X12538695316220

Wilson, E. O. (1984). *Biophilia*. Cambridge, MA: Harvard University Press.

II

NEUROBIOLOGY: APPLYING NEUROSCIENCE TO HUMAN–ANIMAL INTERACTION

4

NEURAL MECHANISMS UNDERLYING HUMAN–ANIMAL INTERACTION: AN EVOLUTIONARY PERSPECTIVE

C. SUE CARTER AND STEPHEN W. PORGES

Social support and positive social interactions can protect and heal (Cacioppo & Patrick, 2008). Thus, the biological causes and consequences of social behavior are important in many fields of science, including neuroscience and medicine. Social interactions have particular value in the face of challenge, especially when the benefits of living socially outweigh those of living alone. Although social behaviors typically involve within-species interactions, these also may be directed toward other species, creating interspecies relationships that are mutually beneficial or symbiotic.

This chapter considers—in the context of evolution—neuroendocrine and autonomic mechanisms underlying the biological consequences of positive social interactions. The neurobiological systems that are implicated in social

We gratefully acknowledge the hard work of many colleagues who provided the original research on which this chapter is based. Much of the research described was supported by the National Institutes of Health, especially the *Eunice Kennedy Shriver* National Institute of Child Health and Human Development. We particularly appreciate the support of Lisa Freund who encouraged the research described here, as well as the writing of this chapter.

http://dx.doi.org/10.1037/14856-006
The Social Neuroscience of Human–Animal Interaction, L. S. Freund, S. McCune, L. Esposito, N. R. Gee, and P. McCardle (Editors)

behavior may help to explain the reported health benefits of human–animal interaction (HAI; e.g., Headey & Grabka, 2007; Headey, Na, & Zheng, 2008). Animal-assisted therapies (AATs), built on HAIs, are increasingly common. Both HAIs and AATs are grounded in the same neural systems that are used to explain the general benefits of social support. Identifying these systems can offer insight into naturally occurring processes through which perceived social support protects or restores human health.

Companion animals can elicit positive emotions and may allow humans to experience a sense of safety, which in turn improves the capacity to regulate both emotional and physiological states. This improved regulation is manifest in better mental and physical health and is often observed as a greater resilience to stressors (Handlin et al., 2011). The behaviors intrinsic to human emotion include dynamic sensations, psychological feelings, and autonomic responses. Social support is most effective when bidirectional and is typically defined by selective social behaviors and social bonds (Carter, 1998), both given and received.

Here we focus primarily on the neurobiology of HAIs under conditions in which these relationships are reciprocal and positive, and thus mutually capable of regulating behavior. However, HAIs present behavioral challenges as well as benefits. Large animals may threaten or harm humans. The coexistence of overlapping neural systems for affiliative support and defensive aggression creates behavioral complexity for the HAI. Thus, awareness of the autonomic and neuroendocrine bases of defensive behaviors can provide insights into situations in which danger or aggression may arise.

BIOLOGICAL PROTOTYPES FOR THE PHYSIOLOGY OF SOCIALITY

The physiological elements that support mammalian maternal behavior are shared with those that underlie social behaviors in general (Carter, 1998, 2014; Carter, Grippo, Pournajafi-Nazarloo, Ruscio, & Porges, 2008). Research in rats revealed that social interactions between the mother and infant are facilitated by oxytocin (Pedersen & Prange, 1979), a hormone released during birth and lactation (Brunton & Russell, 2008). Maternal behavior involves huddling over an infant, which requires social engagement, followed by immobility. In addition, in male prairie voles, the presence of an infant quickly releases oxytocin, which in turn may facilitate or reinforce additional social engagement (Kenkel et al., 2012). In species with selective social behaviors, such as sheep (Keverne, 2006) and socially monogamous prairie voles (Williams, Insel, Harbaugh, & Carter, 1994), oxytocin is a component of the mechanisms that forge social bonds.

However, oxytocin is not the only hormone that can facilitate sociality, nor is its release at birth a requirement for the expression of parental behavior. Mice mutant for the oxytocin gene are capable of birth and some components of maternal behavior; the biochemical systems necessary for birth and maternal behavior are probably redundant (Russell, Leng, & Douglas, 2003). Adoptive parents and alloparents can show high levels of parental behavior. Oxytocin released during these circumstances may be a kind of hormonal "insurance" directing maternal/parental attention toward newborns.

Many other molecules are candidates for the regulation of the causes and consequences of social behavior. As one example, the effects of oxytocin and vasopressin depend on the actions of dopamine, which may help to explain the rewarding effects of maternal behavior (Barrett & Fleming, 2011) and the development of social bonds (Aragona & Wang, 2009).

NEUROENDOCRINOLOGY AND SOCIAL BEHAVIOR

The neural systems responsible for mammalian social cognition and emotion can be influenced by the actions of oxytocin and vasopressin. Both are composed of nine amino acids, and thus are termed *peptides*. The peptide hormones found in mammals evolved from ancient precursors that created the template for modern oxytocin and vasopressin; they may have originated as cellular factors regulating water balance, immune processes, or other defensive mechanisms, which in multicellular organisms are synthesized in particularly high concentrations in the nervous system (sometimes called *neuropeptides*) and have developed the additional capacity to serve as hormones.

The actions of oxytocin and related neuropeptides depend on their capacity to bind to larger protein molecules (*receptors*), generally located on cell membranes. In mammals, peptides and their receptors function together to create a coordinated system, with activities throughout the body (e.g., oxytocin acts on receptors in the brain, peripheral nervous system, uterus, breast, gonads, heart, kidney, and thymus).

Oxytocin serves both as a hormone and a neuromodulator (Landgraf & Neumann, 2004). As a hormone, it is synthesized in the hypothalamus and released into the blood stream at the pituitary, acting on various target organs. As a neuromodulator, it may reach oxytocin receptors throughout the brain by diffusion through the nervous system. Oxytocin can serve as a signaling molecule that influences both behavior and physiology. Through neural pathways in the brainstem, it regulates the autonomic nervous system (ANS) with profound effects on both peripheral organs and emotional states and feelings.

Vasopressin is genetically and structurally related to oxytocin. Research in the socially monogamous prairie vole has implicated oxytocin and vasopressin in sociality, including social contact and pair-bonding, and either can facilitate social engagement with familiar partners. However, selective social behaviors, such as those necessary for pair-bonding (Cho, DeVries, Williams, & Carter, 1999) and responses to infants (Bales, Kim, Lewis-Reese, & Carter, 2004; Bales et al., 2007), appear to require both oxytocin and vasopressin. Because of their structural similarities, these peptides can affect each other's receptors. Untangling the functions of oxytocin and vasopressin has been difficult, but there is compelling evidence that their behavioral actions are not identical (Carter, 2007).

Centrally active vasopressin may be especially critical for the dramatic increase in defensive aggression that supports male pair-bonding (Winslow, Hastings, Carter, Harbaugh, & Insel, 1993) or the paternal defense of offspring (Bosch & Neumann, 2012; Kenkel et al., 2012). Vasopressin (known for increasing blood pressure) also supports physical and probably emotional mobilization (Carter, 1998), as well as other active adaptive and defensive behaviors (Ferris, 2008). It is associated with hyperarousal, vigilance, and aggressive behaviors. As discussed in the following section, knowledge of its functions may be useful in explaining within- and between-species differences in domestication, the defensive aggression that can arise in HAIs, and the negative symptoms, such as depression and anxiety, that emerge when social bonds are disrupted.

EVOLUTION OF MAMMALIAN SOCIAL BEHAVIORS AND HAI

It is likely that the beneficial effects of HAI, as well as other forms of social support, are based in part on older neural systems that emerged before the modern neocortex. This can most easily be understood through an awareness of the evolution of the mammalian nervous system. The nervous system is hierarchically organized. Older neural pathways supporting survival are powerful and can override the more recently evolved cognitive systems. As with other animals, it can be difficult for humans to cortically regulate or understand their visceral reactions and emotional states. The bottom-up design of the mammalian nervous system helps to explain the impact of the comparatively fundamental experiences and feelings that, taken together, can characterize HAI.

The evolved nature of the human central and autonomic nervous systems also helps us understand the origins of mammalian social communication. For example, humans are highly sensitive to the appearance of and vocalizations produced by companion animals; emotional states in both humans and companion animals may, in turn, be regulated by these social signals. A kitten's purring or a dog's barking are common examples of cues that are difficult to ignore, and the evocative visual characteristics of companion animals can elicit emotional feelings in most healthy humans.

In the process of evolution, modern mammals acquired an increasingly sophisticated anatomical and neural capacity to produce and receive complex social cues, including acoustic signals. Among these signals is *prosody* (pitch, rhythm, and tempo), which is fundamental to human language but is also central to social communication in other mammals (Porges, 2011). Prosody can be recognized in domestic animals. Social communication in mammals is based in part on these systems (e.g., some dogs produce sounds that mimic elements of human language). Humans and companion species can be reciprocally sensitive to the types of sounds made by each other and can often use these signals to judge emotional intent and to regulate emotions.

AUTONOMIC NERVOUS SYSTEM, SOCIAL BEHAVIOR, AND EMOTION REGULATION

Of particular importance to social and emotional states is the autonomic core of the mammalian nervous system that manages involuntary functions (e.g., heart, lungs, digestive and immune systems) that emerged long before the modern neocortex. The mammalian ANS also arose as a function of an evolutionary process. Among the evolved functions of the ANS are hierarchical processes that support the high levels of sociality and social communication seen in mammals.

The ANS is functionally divided into sympathetic and parasympathetic components. The sympathetic branches include spinal nerves capable of supporting states of mobilization and some aspects of social engagement but also flight or fight behaviors. As identified by the polyvagal theory (Porges, 2011), the parasympathetic nervous system consists of two vagal components. An older unmyelinated branch of the vagus innervates subdiaphragmatic (i.e., below the diaphragm) organs with effects associated with conservation of energy and reductions in metabolism. In addition, the unmyelinated branch of the vagus slows the heart.

The more recently evolved myelinated vagal system is primarily supradiaphragmatic; it supports oxygenation of the neocortex as well as health, growth, and restoration and is necessary for social engagement. At the level of the mammalian heart, the myelinated vagus is associated with protective neural cardiac rhythms (respiratory sinus arrhythmia), which have been repeatedly associated with health, restoration, and positive social behaviors (Porges, 2011).

ANS AS A NEURAL SUBSTRATE FOR BEHAVIORAL EFFECTS OF OXYTOCIN AND VASOPRESSIN

Peptide hormones, including oxytocin and vasopressin, modulate the functions of the ANS. Oxytocin has complex effects on the sympathetic nervous system that are not well studied, although the acute sympathetic effects of oxytocin may include actions permissive for active social engagement, including play. The chronic actions of oxytocin are more likely to include reductions in sympathetic activation, which over time allow less reactive social interactions such as parenting, at least in comparison to flight or fight reactions (Kenkel et al., 2013). Oxytocin also has actions on the myelinated vagus in humans (Norman et al., 2011) and other social mammals (Grippo, Trahanas, Zimmerman, Porges, & Carter, 2009). The myelinated vagus protects the heart, assuring adequate oxygenation of the neocortex, which in turn is necessary for social interactions and cognition (Porges, 1998). Oxytocin also may directly co-opt the unmyelinated vagus, allowing animals to experience social contact, safely slowing the heart and permitting the "immobility without fear" that characterizes maternal and sexual behaviors, or sitting quietly with a pet. In contrast, "immobility with fear," including freezing or death feigning, is not associated with positive social interactions. "Shut-down" reactions can leave mammals vulnerable to cardiac failures or even death. Oxytocin pathways are protective against a number of these adverse events, including cardiac arrhythmias.

BENEFICIAL EFFECTS OF OXYTOCIN

As described earlier, neuropeptides serve as signaling molecules capable of influencing a variety of neurophysiological functions and at the same time creating coordinated processes associated with growth, health, and restoration. For example, in highly social mammals, such as humans and dogs, oxytocin may bias autonomic reactions and the interpretation of feelings and visceral experiences, allowing emotional states, including those described as a "sense of safety." Alternatively, in the face of perceived threat or danger, a rapid shift in state may facilitate protection of self or a companion.

Social behavior is critically intertwined with stress management. The capacity of companionship and AATs to protect in the face of physical and emotional challenge appears to rely on the biological and molecular substrates that permit the formation of social bonds. Oxytocin may help insure that nonreproductive mammals have access to others in the face of danger. Social bonds can form in response to extreme stressors, especially when survival depends on the presence of another individual; under intense stress, oxytocin is released, leading to the formation of social bonds (Carter, 1998).

Social engagement and support, with an accompanying sense of safety, may be of particular importance to mental health in highly social species (e.g., humans, dogs, prairie voles). In rodents and other social mammals, chronic social isolation is associated with increases in measures of depression, anxiety, and physiological arousal, including changes in basal heart rate and reductions in parasympathetic activity (Grippo et al., 2009). Isolation or other forms of chronic stress also may reduce gene expression for the oxytocin receptor (Pournajafi-Nazarloo et al., 2013), possibly creating insensitivity to the beneficial effects of oxytocin. Concurrently, in female prairie voles isolation is accompanied by an increase in blood levels of oxytocin (Grippo et al., 2009). The autonomic components of voles' responses to chronic isolation are prevented or reversed by chronic treatment with exogenous oxytocin, further implicating oxytocin in social support. Thus in the prairie vole model, isolation-associated elevations in endogenous oxytocin are not sufficient to protect against the autonomic and behavioral consequences of living alone. Whether comparable changes occur in humans remains to be determined, but the presence of a companion, via functional increases in oxytocin or its receptor, could potentially reverse some effects of chronic stress.

As explained, oxytocin can regulate behavioral and emotional reactivity to stress and may help regulate the behavioral and autonomic distress that typically follows separation (Carter, 1998). For example, oxytocin may protect mammals from the "reptile-like" freezing pattern (Porges, 2011). Brainstem regions that influence the ANS have receptors for oxytocin and vasopressin. These same peptides are activated by stressors, and there is increasing evidence that some behavioral consequences of oxytocin and vasopressin are due to their effects on the ANS. Thus conditions that favor activation of oxytocin-regulated processes may enhance resilience in the face of stressful experiences.

BENEFICIAL CONSEQUENCES OF COMPANIONSHIP

Wellness is more than the absence of illness or stress. Social engagement buffers the stress of life. Individuals who have a perceived sense of social support are more resilient in the face of stressors and disease, living longer than those who feel isolated or lonely (Cacioppo & Patrick, 2008). Lesions in various bodily tissues, including the brain, heal more quickly in animals that are living socially versus in isolation (Karelina & DeVries, 2011). Remarkably, the same hormones and brain areas that serve the body's survival demands also permit adaptation to an ever-changing social and physical environment.

The protective effects of positive sociality appear to rely on a concoction of naturally occurring molecules with diverse actions throughout the body.

As one example, oxytocin may literally heal a damaged heart, promoting growth and restoration. Oxytocin receptors are expressed in the heart, and precursors for the oxytocin peptide appear to be critical to the development of the fetal heart (Danalache, Gutkowska, Slusarz, Berezowska, & Jankowski, 2010). Oxytocin has protective effects in part through the capacity to convert undifferentiated stem cells into cardiomyocytes and may facilitate adult neurogenesis, especially after a stressful experience. It also has direct anti-inflammatory and antioxidant properties that can be detected in in vitro models of atherosclerosis (Szeto et al., 2008).

Oxytocin, in conjunction with vasopressin, plays a dynamic role in the regulation of the ANS (Grippo et al., 2009; Porges, 2011). Oxytocin and vasopressin receptors also are found in brainstem regions that regulate heart rate. Both peptides modulate the sympathetic system, but it is probably oxytocin that normally restrains the overreactivity of the cardiovascular system in the face of extreme or chronic stress, as it has direct actions on the parasympathetic component of the ANS, permitting and promoting some of the benefits of social engagement.

IS OXYTOCIN AN ANTISTRESS HORMONE?

Initial research on the social consequences of oxytocin emphasized its capacity to downregulate the hypothalamic–pituitary–adrenal (HPA) axis (Carter, 1998). As data have accumulated, these overly simplistic perspectives have been replaced by more sophisticated views of the mechanisms through which the oxytocin or social support can play an adaptive role in the response to a challenge. In addition, we have become increasingly aware of the importance of the ANS in the regulation of responses to challenge (Porges, 2011). As described earlier, the behavioral and autonomic consequences of exposure to oxytocin, vasopressin, or other hormones of the HPA axis are time- and context-dependent. Positive experiences may in some cases release oxytocin but are not the only mechanisms for stimulating oxytocin's release. Among the most reliable methods to release oxytocin are intense stressors, including birth. Under conditions of high arousal, the most immediate effects of exposure to oxytocin may be a transient activation of the HPA axis and sympathetic nervous system. Thus, interpretations of blood levels of hormones are not simple. High levels of oxytocin may be detected following both positive and negative experiences. The biological consequences of oxytocin levels may be best understood in the context of the status of the oxytocin (and vasopressin) receptors, but at present functional measures of receptors are not easily assessed, and we are limited to estimations of receptor activity based on measures of genetic and epigenetic markers.

Timing of hormone measurements and knowledge of individual differences and context are critical. Oxytocin has complex, dynamic effects on the HPA axis and the ANS, with potentially compensatory consequences for coping with challenge. Under some conditions, oxytocin can inhibit the release of glucocorticoids (a class of steroid hormones; cortisol is the most important human glucocorticoid). As with oxytocin, the effects of glucocorticoids also are time dependent. People given cortisol-like drugs are initially euphoric, although chronic exposure to glucocorticoids is associated with fatigue and depression (Judd et al., 2014).

BETWEEN- AND WITHIN-SPECIES VARIATIONS IN SOCIAL BEHAVIOR

Over the millennia, biological substrates emerged that permitted the expression of the modern mammalian social engagement and permitted the experiences that humans call *emotions* (Donaldson & Young, 2008). Of particular relevance to HAIs are mechanisms for positive sociality and social bonds. These neural and anatomical systems are the underpinnings for contemporary social behaviors, including those observed in humans and most of their preferred companion animals. The evolved traits that allow a species to become a domestic species include the capacity to interact with humans over an extended period of time and in some cases to form selective social bonds.

Domestic animals have been bred to serve humans as companions, pets, and guards. The ancestors of species that could be domesticated were presumably also social to some extent. For example, modern dogs are believed to have arisen from wolves; both share with humans the capacity to develop social bonds. It is likely that shared physiological traits and habitats allowed the development of particularly strong ties between humans and dogs. Dogs also may serve to alert or protect against intruders, culminating with their well-established designation as "man's best friend" (Olmert, 2009).

SOCIAL ENGAGEMENT CAN ALLOW A SENSE OF SAFETY AND IMPROVE STATE REGULATION

The capacity to socially engage, giving and receiving high levels of social behavior and especially selective social relationships, is of particular relevance to HAI. Humans, who may experience an enhanced sense of safety in the presence of human companions, can experience a similar sense of emotional safety with companion animals or pets. In fact, the affection of a pet may seem more unconditional and may be less complex to experience and return than that of human companions.

There is increasing evidence for health benefits of social support, especially in the face of extreme challenges or trauma (Olff, 2012). The neurophysiological consequences of social behavior can be profound, probably translating into a sense of biological and emotional safety. The benefits of social support are most readily detected when absent: Involuntary separation from a loved one, isolation, or a perceived sense of loneliness are associated with a host of negative consequences, including increases in depression and vulnerability to various diseases (Cacioppo & Patrick, 2008). Death of a pet can create a similar sense of loss. As with other positive influences on physiology and behavior, the benefits of HAI may be most easily detected when a need state exists, including in the face of external stress. Companion animals also can be of particular value during stressful periods, including bereavement or emotional loss.

ANIMAL-ASSISTED THERAPIES

Systematic efforts to access the value of AATs, most often using dogs, cats, or birds, have been reported in dozens of studies and across the lifespan (reviewed in Beetz, Uvnäs-Moberg, Julius, & Kotrschal, 2012). Among the commonly measured physiological changes described as a function of HAIs are reductions in heart rate, blood pressure, cortisol, and catecholamines. These changes can appear in healthy humans but may be most obvious in individuals faced with acute or chronic stressors or illness; benefits of HAIs have been reported for cardiovascular diseases, cancer, dementias, and various mental illnesses, including depression, schizophrenia, and autism. The diversity of diseases that gain benefit from AATs suggests that the biology of these is based on fundamental physiological processes with wide consequences.

CHALLENGES IN STUDYING OXYTOCIN
AND VASOPRESSIN IN HAI

The neurobiological processes through which HAIs are beneficial to either the humans or their animal companions deserve further study. Clues to the underlying mechanisms are emerging from other behavioral models (e.g., maternal behavior) and from studies of socially monogamous rodents, which like humans and dogs have the capacity to show high levels of sociality and form long-lasting pair-bonds (Carter, DeVries, & Getz, 1995). These studies have repeatedly implicated oxytocin and vasopressin in social behavior, both during early life and in adulthood (Carter, Boone, Pournajafi-Nazarloo, & Bales, 2009; Feldman, 2012; Meyer-Lindenberg, Domes, Kirsch, & Heinrichs, 2011).

Studies examining the specific neural mechanisms associated with HAIs have begun to focus on peptides implicated in social behavior (e.g., endocrine changes as a function of interactions with dogs include increases in oxytocin; Handlin et al., 2011; Nagasawa, Kikusui, Onaka, & Ohta, 2009; Odendaal & Meintjes, 2003). Among the challenges in studying possible changes associated with exposure to an animal companion is the fact that changes may be small or transient; both sensitive measures and frequent sampling are required. The status of the oxytocin receptor may alter the consequences of exposure to either endogenous or exogenous hormones, although at present this can only be estimated by measures of genetic and epigenetic markers.

Differences are to be expected among individuals, species, and breeds but at present are not well understood. Paradigms that can be used for studies of HAI depend on the strength of the relationship between the companion and human (Rehn, Handlin, Uvnäs-Moberg, & Keeling, 2014). Depending on the circumstance or individuals involved, changes may occur in one or both members of the dyad. In one study, levels of urinary oxytocin were measured as a function of human–pet interactions (Nagasawa et al., 2009). In comparing owners whose dog showed "long" versus "short" gaze toward the owner, only those owners with long-gaze dogs showed increases in oxytocin following an interaction with their pets (see Figure 4.1). In another study, repeated dog and human blood samples were taken immediately following interactions; changes were apparent within 1 to 3 minutes, and although

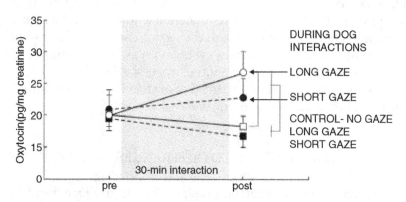

Figure 4.1. Oxytocin (urinary) in pet owners increased between before (pre) and after (post) typical interactions with their dogs when the dogs were long-gaze-toward owner dogs. From "Dog's Gaze at Its Owner Increases Owner's Urinary Oxytocin During Social Interaction," by M. Nagasawa, T. Kikusui, T. Onaka, and M. Ohta, 2009, *Hormones and Behavior, 55*, p. 438. Copyright 2008 by Elsevier, Inc. Adapted with permission.

increases in oxytocin were statistically significant in humans, it was dogs that showed the most marked increase (Handlin et al., 2011). In addition, the nature of the interaction between a dog and its partner can affect the oxytocin response. Physical (versus nonphysical) contact with a familiar partner has been associated with a longer lasting increase in oxytocin in dogs (Rehn et al., 2014). Finally, there is recent evidence that exogenous oxytocin, given to dogs as an intranasal spray, can influence social behavior. Oxytocin-treated animals were more likely to show positive social interactions toward both their owners and other dogs (Romero, Nagasawa, Mogi, Hasegawa, & Kikusui, 2014).

Additional evidence for the power of companion animals to influence humans comes from recent studies in dogs showing that exogenous oxytocin given to a dog not only increased gazing by the dog toward its owner, but also was associated with a release of oxytocin in the owner (see Figure 4.1; Nagasawa et al., 2015). It is interesting to note that these effects were not seen in interactions between domesticated wolves and their owners. Results from these studies also have been used to support the argument that bonds between humans and dogs represent an example of "coevolution" (Nagasawa et al., 2015) or "evolutionary convergence" in which mechanisms typically used for within-species attachment are "hijacked," permitting the emergence of cross-species bonds (MacLean & Hare, 2015).

Taken together, these studies suggest that under some conditions, HAIs may release oxytocin. The release of oxytocin by social behaviors can be transitory or long lasting and can be affected by the relationship between dogs and their social partners. Furthermore, the capacity of oxytocin to elicit social behavior has been documented using exogenous oxytocin. Although changes have been measured in a variety of other chemicals during HAIs, the degree to which these are specific to social interactions, rather than reflecting more general changes associated with the testing conditions, have not been fully addressed (e.g., circumstances experienced as arousing may be either negative or positive). Thus, at present the strongest available evidence comes from dogs and supports the hypothesis that oxytocin may play a central role in the benefits of human–dog interaction.

SOCIAL BEHAVIORS, BREED DIFFERENCES, AND PEPTIDES

Dog breeds have been carefully sculpted by selective breeding. In some cases the breeding objective was to create an animal that would protect people or assist in hunting game, whereas in others, animals were bred as companions to help humans regulate their emotions. It is plausible that breed differences are related to differences in oxytocin and vasopressin. For example, highly social breeds, especially those that maintain puppy-like behavioral and physical features into adulthood (neoteny), may be expressing oxytocin-dependent traits.

A major issue in HAIs is understanding conditions under which either animals or their human caretakers may express aggression toward each other. The hormones that support guarding of a partner in particular are part of a delicately balanced system. Aggression is most often expressed in the face of challenge and toward perceived intruders but can misfire, appearing in unintended contexts or with undesired consequences. Other animals, or even friends and family, may become the object of aggression. The vasopressin system can support mobilization, defensive aggression, and anxiety (Ferris, 2008; Zhang et al., 2012). Breeds of dogs that show guarding behaviors may have been inadvertently bred for high levels of central vasopressin. A strong and selective social bond might be associated with a tendency to show defensive aggression, usually toward strangers, by some companion dogs.

To our knowledge, the relationships between oxytocin or ANS features and species and breed differences in sociality have not been systematically examined. However, knowledge of the role of the ANS and neuropeptides, such as oxytocin and vasopressin and their receptors, in the social behavior of pets or other domestic species could be useful to the creation of breeds of companion or domestic animals with a desired behavioral profile conducive to social bonding. Such knowledge might also be useful in selecting rescued animals with the highest chance of being successfully adopted.

SUMMARY

Evolution is thrifty, resculpting and repurposing molecules and neural systems. The modern nervous system evolved using chemicals with multiple sites of action, many of which were retained as additional functions emerged. They are components of integrated neural networks, coordinating sociality with other bodily processes. As with birth, social engagement is so critical that the systems regulating it may be redundant. It is likely that the neural systems and molecules that have been identified as underpinnings of social behaviors in rodents and humans are also critical to HAI. Domestic animals were derived from social ancestors and rely on neural pathways similar to those in humans. It would be possible, taking advantage of existing within-species or breed differences, to examine whether artificial selection and domestication act indirectly through manipulations of peptides. Oxytocin might be associated with a general tendency toward sociality and possibly neoteny. Both oxytocin and vasopressin, probably acting via the myelinated supradiaphragmatic vagus, may be necessary to produce animals showing strong selective social bonds, including toward human caretakers. However, vasopressin, which within the central nervous system is androgen dependent (Carter, 2007), is more likely to be associated with defensive or protective traits, expected to arise in adulthood, and to be context and gender dependent. Finally, the growing literature on HAI offers a novel opportunity to gain a deeper understanding of the evolutionary and neurobiological basis of mammalian sociality.

REFERENCES

Aragona, B. J., & Wang, Z. (2009). Dopamine regulation of social choice in a monogamous rodent species. *Frontiers in Behavioral Neuroscience*, 3, 15. http://dx.doi.org/10.3389/neuro.08.015.2009

Bales, K. L., Kim, A. J., Lewis-Reese, A. D., & Carter, C. S. (2004). Both oxytocin and vasopressin may influence alloparental behavior in male prairie voles. *Hormones and Behavior*, 45, 354–361. http://dx.doi.org/10.1016/j.yhbeh.2004.01.004

Bales, K. L., van Westerhuyzen, J. A., Lewis-Reese, A. D., Grotte, N. D., Lanter, J. A., & Carter, C. S. (2007). Oxytocin has dose-dependent developmental effects on pair-bonding and alloparental care in female prairie voles. *Hormones and Behavior*, 52, 274–279. http://dx.doi.org/10.1016/j.yhbeh.2007.05.004

Barrett, J., & Fleming, A. S. (2011). All mothers are not created equal: Neural and psychobiological perspective on mothering and the importance of individual differences. *Journal of Child Psychology and Psychiatry*, 52, 368–397. http://dx.doi.org/10.1111/j.1469-7610.2010.02306.x

Beetz, A., Uvnäs-Moberg, K., Julius, H., & Kotrschal, K. (2012). Psychosocial and psychophysiological effects of human-animal interactions: The possible role of oxytocin. *Frontiers in Psychology*, 3, 234. http://dx.doi.org/10.3389/fpsyg.2012.00234

Bosch, O. J., & Neumann, I. D. (2012). Both oxytocin and vasopressin are mediators of maternal care and aggression in rodents: From central release to sites of action. *Hormones and Behavior*, 61, 293–303. http://dx.doi.org/10.1016/j.yhbeh.2011.11.002

Brunton, P. J., & Russell, J. A. (2008). The expectant brain: Adapting for motherhood. *Nature Reviews Neuroscience*, 9, 11–25. http://dx.doi.org/10.1038/nrn2280

Cacioppo, J. T., & Patrick, W. (2008). *Loneliness: Human nature and the need for social connection*. New York, NY: Norton.

Carter, C. S. (1998). Neuroendocrine perspectives on social attachment and love. *Psychoneuroendocrinology*, 23, 779–818. http://dx.doi.org/10.1016/S0306-4530(98)00055-9

Carter, C. S. (2007). Sex differences in oxytocin and vasopressin: Implications for autism spectrum disorders? *Behavioural Brain Research*, 176, 170–186. http://dx.doi.org/10.1016/j.bbr.2006.08.025

Carter, C. S. (2014). Oxytocin pathways and the evolution of human behavior. *Annual Review of Psychology*, 65, 17–39. http://dx.doi.org/10.1146/annurev-psych-010213-115110

Carter, C. S., Boone, E. M., Pournajafi-Nazarloo, H., & Bales, K. L. (2009). Consequences of early experiences and exposure to oxytocin and vasopressin are sexually dimorphic. *Developmental Neuroscience, 31*, 332–341. http://dx.doi. org/10.1159/000216544

Carter, C. S., DeVries, A. C., & Getz, L. L. (1995). Physiological substrates of mammalian monogamy: The prairie vole model. *Neuroscience and Biobehavioral Reviews, 19*, 303–314. http://dx.doi.org/10.1016/0149-7634(94)00070-H

Carter, C. S., Grippo, A. J., Pournajafi-Nazarloo, H., Ruscio, M. G., & Porges, S. W. (2008). Oxytocin, vasopressin and sociality. *Progress in Brain Research, 170*, 331–336. http://dx.doi.org/10.1016/S0079-6123(08)00427-5

Cho, M. M., DeVries, A. C., Williams, J. R., & Carter, C. S. (1999). The effects of oxytocin and vasopressin on partner preferences in male and female prairie voles (*Microtus ochrogaster*). *Behavioral Neuroscience, 113*, 1071–1079. http:// dx.doi.org/10.1037/0735-7044.113.5.1071

Danalache, B. A., Gutkowska, J., Slusarz, M. J., Berezowska, I., & Jankowski, M. (2010). Oxytocin-Gly-Lys-Arg: A novel cardiomyogenic peptide. *PLoS ONE, 5*(10), e13643. http://dx.doi.org/10.1371/journal.pone.0013643

Donaldson, Z. R., & Young, L. J. (2008). Oxytocin, vasopressin, and the neurogenetics of sociality. *Science, 322*, 900–904. http://dx.doi.org/10.1126/science.1158668

Feldman, R. (2012). Oxytocin and social affiliation in humans. *Hormones and Behavior, 61*, 380–391. http://dx.doi.org/10.1016/j.yhbeh.2012.01.008

Ferris, C. F. (2008). Functional magnetic resonance imaging and the neurobiology of vasopressin and oxytocin. *Progress in Brain Research, 170*, 305–320. http:// dx.doi.org/10.1016/S0079-6123(08)00425-1

Grippo, A. J., Trahanas, D. M., Zimmerman, R. R., II, Porges, S. W., & Carter, C. S. (2009). Oxytocin protects against negative behavioral and autonomic consequences of long-term social isolation. *Psychoneuroendocrinology, 34*, 1542–1553. http://dx.doi.org/10.1016/j.psyneuen.2009.05.017

Handlin, L., Hydbring-Sandberg, E., Nilsson, A., Ejdebäck, M., Jansson, A., & Uvnäs-Moberg, K. (2011). Short-term interaction between dogs and their owners—effects on oxytocin, cortisol, insulin and heart rate—an exploratory study. *Anthrozoös, 24*, 301–315. http://dx.doi.org/10.2752/175303711X13045914865385

Headey, B., & Grabka, M. M. (2007). Pets and human health in Germany and Australia: National Longitudinal Results. *Social Indicators Research, 80*, 297–311. http://dx.doi.org/10.1007/s11205-005-5072-z

Headey, B., Na, F., & Zheng, R. (2008). Pet dogs benefit owners' health: A "natural experiment" in China. *Social Indicators Research, 87*, 481–493. http://dx.doi. org/10.1007/s11205-007-9142-2

Judd, L. L., Schettler, P. J., Brown, E. S., Wolkowitz, O. M., Sternberg, E. M., Bender, B. G., . . . Singh, G. (2014). Adverse consequences of glucocorticoid medication: Psychological, cognitive, and behavioral effects. *The American Journal of Psychiatry, 171*, 1045–1051. http://dx.doi.org/10.1176/appi.ajp.2014.13091264

Karelina, K., & DeVries, A. C. (2011). Modeling social influences on human health. *Psychosomatic Medicine*, *73*, 67–74. http://dx.doi.org/10.1097/PSY.0b013e3182002116

Kenkel, W. M., Paredes, J., Yee, J. R., Pournajafi-Nazarloo, H., Bales, K. L., & Carter, C. S. (2012). Neuroendocrine and behavioural responses to exposure to an infant in male prairie voles. *Journal of Neuroendocrinology*, *24*, 874–886. http://dx.doi.org/10.1111/j.1365-2826.2012.02301.x

Kenkel, W. M., Paredes, J., Lewis, G. F., Yee, J. R., Pournajafi-Nazarloo, H., Grippo, A. J., . . . Carter, C. S. (2013). Autonomic substrates of the response to pups in male prairie voles. *PLoS ONE*, *8*(8), e69965. http://dx.doi.org/10.1371/journal.pone.0069965

Keverne, E. B. (2006). Neurobiological and molecular approaches to attachment and bonding. In C. S. Carter, L. Ahnert, K. E. Grossman, S. B. Hrdy, M. E. Lamb, S. W. Porges, & N. Saschser (Eds.), *Attachment and bonding: A new synthesis* (pp. 101–118). Cambridge, MA: MIT Press.

Landgraf, R., & Neumann, I. D. (2004). Vasopressin and oxytocin release within the brain: A dynamic concept of multiple and variable modes of neuropeptide communication. *Frontiers in Neuroendocrinology*, *25*, 150–176. http://dx.doi.org/10.1016/j.yfrne.2004.05.001

MacLean, E. L., & Hare, B. (2015). Dogs hijack the human bonding pathway. *Science*, *348*, 280–281. http://dx.doi.org/10.1126/science.aab1200

Meyer-Lindenberg, A., Domes, G., Kirsch, P., & Heinrichs, M. (2011). Oxytocin and vasopressin in the human brain: Social neuropeptides for translational medicine. *Nature Reviews Neuroscience*, *12*, 524–538. http://dx.doi.org/10.1038/nrn3044

Nagasawa, M., Kikusui, T., Onaka, T., & Ohta, M. (2009). Dog's gaze at its owner increases owner's urinary oxytocin during social interaction. *Hormones and Behavior*, *55*, 434–441. http://dx.doi.org/10.1016/j.yhbeh.2008.12.002

Nagasawa, M., Mitsui, S., En, S., Ohtani, N., Ohta, M., Sakuma, Y., . . . Kikusui, T. (2015). Oxytocin-gaze positive loop and the coevolution of human-dog bonds. *Science*, *348*, 333–336. http://dx.doi.org/10.1126/science.1261022

Norman, G. J., Cacioppo, J. T., Morris, J. S., Malarkey, W. B., Berntson, G. G., & Devries, A. C. (2011). Oxytocin increases autonomic cardiac control: Moderation by loneliness. *Biological Psychology*, *86*, 174–180. http://dx.doi.org/10.1016/j.biopsycho.2010.11.006

Odendaal, J. S., & Meintjes, R. A. (2003). Neurophysiological correlates of affiliative behaviour between humans and dogs. *Veterinary Journal*, *165*, 296–301. http://dx.doi.org/10.1016/S1090-0233(02)00237-X

Olff, M. (2012). Bonding after trauma: On the role of social support and the oxytocin system in traumatic stress. *European Journal of Psychotraumatology*, *3*. http://dx.doi.org/10.3402/ejpt.v3i0.18597

Olmert, M. D. (2009). *Made for each other: The biology of the human-animal bond.* Cambridge, MA: Da Capo Press.

Pedersen, C. A., & Prange, A. J., Jr. (1979). Induction of maternal behavior in virgin rats after intracerebroventricular administration of oxytocin. *Proceedings of the National Academy of Sciences, USA, 76,* 6661–6665. http://dx.doi.org/10.1073/pnas.76.12.6661

Porges, S. W. (1998). Love: An emergent property of the mammalian autonomic nervous system. *Psychoneuroendocrinology, 23,* 837–861. http://dx.doi.org/10.1016/S0306-4530(98)00057-2

Porges, S. W. (2011). *The polyvagal theory: Neurophysiological foundations of emotions, attachment, communication and self-regulation.* New York, NY: Norton.

Pournajafi-Nazarloo, H., Kenkel, W., Mohsenpour, S. R., Sanzenbacher, L., Saadat, H., Partoo, L., . . . Carter, C. S. (2013). Exposure to chronic isolation modulates receptors mRNAs for oxytocin and vasopressin in the hypothalamus and heart. *Peptides, 43,* 20–26. http://dx.doi.org/10.1016/j.peptides.2013.02.007

Rehn, T., Handlin, L., Uvnäs-Moberg, K., & Keeling, L. J. (2014). Dogs' endocrine and behavioural responses at reunion are affected by how the human initiates contact. *Physiology & Behavior, 124,* 45–53. http://dx.doi.org/10.1016/j.physbeh.2013.10.009

Romero, T., Nagasawa, M., Mogi, K., Hasegawa, T., & Kikusui, T. (2014). Oxytocin promotes social bonding in dogs. *Proceedings of the National Academy of Sciences, USA, 111,* 9085–9090. http://dx.doi.org/10.1073/pnas.1322868111

Russell, J. A., Leng, G., & Douglas, A. J. (2003). The magnocellular oxytocin system, the fount of maternity: Adaptations in pregnancy. *Frontiers in Neuroendocrinology, 24,* 27–61. http://dx.doi.org/10.1016/S0091-3022(02)00104-8

Szeto, A., Nation, D. A., Mendez, A. J., Dominguez-Bendala, J., Brooks, L. G., Schneiderman, N., & McCabe, P. M. (2008). Oxytocin attenuates NADPH-dependent superoxide activity and IL-6 secretion in macrophages and vascular cells. *American Journal of Physiology, Endocrinology and Metabolism, 295,* E1495–E1501. http://dx.doi.org/10.1152/ajpendo.90718.2008

Williams, J. R., Insel, T. R., Harbaugh, C. R., & Carter, C. S. (1994). Oxytocin administered centrally facilitates formation of a partner preference in female prairie voles (*Microtus ochrogaster*). *Journal of Neuroendocrinology, 6,* 247–250. http://dx.doi.org/10.1111/j.1365-2826.1994.tb00579.x

Winslow, J. T., Hastings, N., Carter, C. S., Harbaugh, C. R., & Insel, T. R. (1993). A role for central vasopressin in pair bonding in monogamous prairie voles. *Nature, 365,* 545–548. http://dx.doi.org/10.1038/365545a0

Zhang, L., Hernández, V. S., Liu, B., Medina, M. P., Nava-Kopp, A. T., Irles, C., & Morales, M. (2012). Hypothalamic vasopressin system regulation by maternal separation: Its impact on anxiety in rats. *Neuroscience, 215,* 135–148. http://dx.doi.org/10.1016/j.neuroscience.2012.03.046

5

AFFILIATION IN HUMAN–ANIMAL INTERACTION

ANDREA BEETZ AND KAREN BALES

Interactions and relationships of humans with animals can take many forms, from having working partnerships and close social relationships to, in the worst case, abusing animal companions. Modern Western societies are profoundly "pet-ified"; many owners regard their companion animals as friends and family members and report a strong emotional attachment to them. But why are humans, living in a modern and technically dominated world, so fascinated with interactions with animals? On what common base can humans and animals develop such social interspecies relationships, and what are the biological mechanisms underlying the many positive effects of human–animal interaction (HAI)? The other chapters in this section address these questions of neurobiological and physiological mechanisms. While referencing these previous chapters, we add a short overview of some major concepts that are particularly related to the evolutionary basis and psychobiological mechanisms of HAI: biophilia, attachment theory and bonding, social

http://dx.doi.org/10.1037/14856-007
The Social Neuroscience of Human–Animal Interaction, L. S. Freund, S. McCune, L. Esposito, N. R. Gee, and P. McCardle (Editors)

support, oxytocin and stress systems, classical conditioning, and distraction affects. Finally, we provide our thoughts for future research directions and implications for the practice of animal-assisted intervention (AAI).

HUMAN BIOPHILIA

Infants, and older children even more so, show more interest in and attention to animals and living beings than to nonliving objects (DeLoache, Pickard, & LoBue, 2011). Although individual differences in this preference increase in the course of ontogeny, a majority of adults still show some interest in interacting with animals and nature. This has prompted the conclusion that humans are *biophilic* (Kellert & Wilson, 1993; Wilson, 1984). This is not surprising, because during their evolutionary history, humans have always lived in close contact with nature. Therefore, paying attention to animals in their surroundings was certainly adaptive for the survival and fitness of early humans—for example, for avoiding predators but also for hunting or taming wildlife (Serpell, 1986; Wilson, 1984). Furthermore, the behaviors of animals may have served as clues for whether a certain environment was safe or not. Via subconscious perception, a relaxed, resting animal seems to relax humans and in a contagious way to down-regulate their stress systems (Julius, Beetz, Kotrschal, Turner, & Uvnäs-Moberg, 2013). This physiological as well as psychological calming effect and feeling of security via the presence of relaxed animals has been described as the *biophilia effect* (Julius et al., 2013) and may also be the basis for the "neuroception of safety," a phrase coined by Porges in the framing of his polyvagal theory (see Chapter 4, this volume).

POSITIVE EFFECTS OF HAI ON HUMANS

Today, sound research with controlled designs documents the following effects of HAI, frequently in comparison with interactions between humans only (for a review and references, see Beetz, Uvnäs-Moberg, Julius, & Kotrschal, 2012):

- In the company of a friendly animal (especially a friendly-looking dog), humans receive increased positive social attention from others. This effect has been documented, for example, for teachers in a classroom with a dog, for strangers passing by, and for persons in wheelchairs. An explanation could be that the biophilic interest in the animal overrides some human societal norms or habits of ignoring others, especially when paying attention could lead to feelings of uncertainty (e.g., due to a person's disability). Furthermore, the presence of animals increases social interactions, including verbal communication between humans in casual encounters and in AAIs with humans of all ages, with and without physical or mental disorders. This is commonly referred to as the *social catalyst effect* of animals.

- Animals seem to promote trust toward the accompanied person or her or his perceived trustworthiness, be it a psychotherapist or a stranger in the street.
- Interacting with animals, in particular, dogs, may reduce depression and facilitate a positive mood. This has been investigated mainly in children and in elderly clients with and without mental or physical disorders.
- Animals may contribute to reducing anxiety and to promoting calmness, especially when the person is expecting or experiencing a stressful situation, such as an unpleasant medical treatment or a fear-inducing health condition.
- Among the most frequently investigated effects of HAI is the buffering of the human stress systems. Several studies using endocrinological parameters have documented that interaction with a friendly dog can decrease cortisol levels in plasma and saliva in children and adults as well as levels of epinephrine and norepinephrine. In addition, cardiovascular parameters indicating the activation of stress systems can be positively influenced by HAIs. In the absence of a specific stressor, children and adults had lower blood pressure when interacting with a dog in comparison with reading or interacting with a person. Finally, heart rate and heart-rate variability may be positively affected by interacting with dogs. Similar effects have been observed in studies with children and adults in specific stressful situations.
- Potentially positive effects of HAI on empathy, aggression, concentration, and pain management have been documented.
- Companion animal ownership has been associated with general health benefits. Dog and cat owners make fewer doctor visits and are less likely to take medication for sleeping problems or heart problems than nonpet-owning control populations, and dog owners report better fitness and health and have a higher likelihood of long-term survival after a myocardial infarction.

Supporting research for the above effects points to some important factors. Most of the effects, in particular, antistress effects, seem to be more pronounced when participants interact with their own pets in comparison to unfamiliar animals (Odendaal, 2000; Odendaal & Meintjes, 2003), and the animals seem to be even more effective than close friends or partners (Allen, Blascovich, & Mendes, 2002; Allen, Blascovich, Tomaka, & Kelsey, 1991). Furthermore, the quality of the interaction, for example, the amount of physical contact, seems to play a role (Beetz et al., 2011).

However, results like these are generated within an optimal setting, that is, with voluntary participants who probably place high value on their relationship to their pets. It seems likely that a positive-to-neutral attitude toward the involved animal species and no fear of the involved animals are preconditions for the reported HAI effects. Similarly, the involved animals are usually selected for obedience, calmness, friendliness, and health and have been well socialized with humans.

PSYCHOLOGICAL AND PHYSIOLOGICAL MECHANISMS EXPLAINING THE EFFECTS OF HAI AND HUMAN–ANIMAL RELATIONSHIPS

The following theoretical and mechanistic frameworks seem useful for the explanation of the positive effects of HAI discussed previously, for the specific benefit of companion animals in comparison with other humans, and toward explaining why it is possible to establish rich social relationships between humans and their pets: the oxytocin system and other theory frames rooted in physiology, such as the polyvagal theory, bonding and attachment theory, social support, and classical conditioning and distraction effects. However, this list is not an all-inclusive account of theoretical approaches in HAI, and these theory frames are by no means mutually exclusive.

Activation of the Oxytocin System

Due to many parallels in the effects of HAI and the known effects linked with the hormone oxytocin, it seems plausible to assume that the activation of the oxytocin system might play a significant role in most HAIs and that it might represent one of the underlying mechanisms of HAI effects (Beetz, Uvnäs-Moberg, et al., 2012; Julius et al., 2013).

Oxytocin (OT) is a peptide hormone that is released into the brain and circulatory system from the hypothalamus by sensory stimulation, for example, during sex, breastfeeding, or stroking (for an overview, see Beetz et al., 2011; Insel, 2010; Uvnäs-Moberg, 2003; see also Chapter 4, this volume). OT is involved in many physiological functions, including milk letdown, uterine contractions during labor, and orgasm (Zingg, 2002). OT can decrease stress (especially in regard to social stressors; Heinrichs, Baumgartner, Kirschbaum, & Ehlert, 2003), including decreasing cortisol levels (Cardoso, Ellenbogen, Orlando, Bacon, & Joober, 2013; Heinrichs et al., 2003) and blood pressure (Petersson & Uvnäs-Moberg, 2007, 2008). OT also decreases anxiety and depression (Neumann & Landgraf, 2012) and increases the pain threshold (Bodnar, Nilaver, Wallace, Badillo-Martinez, & Zimmerman, 1984; Yang et al., 2011).

In addition to the effects OT has on stress and anxiety, which might facilitate social interactions, OT itself may play a direct role in human–animal bonding. OT is involved in maternal–infant bonding in many species, as well as paternal behavior in those species where it has been examined (Bales, Boone, Epperson, Hoffman, & Carter, 2011; Bales, Kim, Lewis-Reese, & Carter, 2004; Feldman, Gordon, & Zagoory-Sharon, 2011; Kenkel et al., 2012; Pedersen & Prange, 1979). In socially monogamous animals, such as prairie voles (*Microtus ochrogaster*), OT is involved in the formation of adult heterosexual pair-bonds (particularly in females, although it plays a role in both sexes; Cho, DeVries, Williams, & Carter, 1999; Ross et al., 2009; Williams, Insel, Harbaugh, & Carter, 1994). In human studies, exogenously administered OT stimulates social interaction, including eye contact, trust, positive self-perception, generosity, and social skills (Guastella & MacLeod, 2012; van IJzendoorn & Bakermans-Kranenburg, 2012). Caregiving interactions in humans, including contact, synchrony between infants and parents, and social vocalizations, can also lead to increases of OT in saliva (Feldman et al., 2011; Seltzer, Ziegler, & Pollak, 2010).

Several studies have documented effects of HAI on the OT system in humans. A significant increase of OT in plasma in both humans and dogs, after 5 to 24 minutes of stroking the animals, was reported by Odendaal (2000; Odendaal & Meintjes, 2003). The effect was even stronger when owners interacted with their own dog in comparison with an unfamiliar dog. Handlin et al. (2011) found an increase in plasma OT after female owners had stroked and talked to their dogs for 3 minutes. Also, interaction and eye contact between highly attached owners and their own dogs was shown to be associated with an increase in OT, measured in urine (Nagasawa, Kikusui, Onaka, & Ohta, 2009).

It is therefore possible that OT is involved in HAI via a number of pathways, including both the reduction of stress and anxiety and an increase in affiliative bonding. Finally, OT may exert some of its actions in HAI indirectly via up-regulating the parasympathetic system (see this chapter's section on polyvagal theory).

Attachment and Social Support

Although OT provides a physiological mechanism for attachment, attachment theory itself provides a comprehensive psychological explanatory basis for human–animal relationships and positive effects of HAI. Even though Ainsworth (1963) and Bowlby (1969), the founders of attachment theory, only addressed human–human relationships, this concept may explain why many humans profit more from social support by an animal than from a friendly human when under stress (Beetz, Julius, Turner, & Kotrschal, 2012; Beetz et al., 2011; Julius et al., 2013). The concept of attachment can be used to capture the quality of the human–animal relationship and to understand the goals of social behaviors within a human–animal dyad.

Via experiences with their primary caregivers, children develop their specific attachment representation during their first year of life (Bowlby, 1969). Attachment is a behavioral system that has as its primary function to establish and maintain proximity to the caregiver, thus to ensure protection and caregiving for the child (George & Solomon, 2008; Marvin & Britner, 2008). The attachment behavioral system also regulates stress. Attachment is mentally represented and regulated via so-called internal working models (Bretherton & Munholland, 2008), which affect how children respond to their primary caregivers on the basis of previous experiences. An effective caregiver serves as a secure base for exploration of the environment and as a safe haven when under stress (Bowlby, 1969). Secure attachment develops in interactions with sensitive and reliable caregivers (Ainsworth, Bell, & Stayton, 1971), whereas insecure (avoidant or ambivalent) or disorganized attachment as secondary adaptive strategies results primarily from suboptimal caregiving. Securely attached infants seek to establish proximity to the caregiver when under stress and can use that effectively for stress regulation (Ainsworth et al., 1971; Ainsworth & Wittig, 1969). In contrast, children with insecure or disorganized attachment profit significantly less from contact with the caregiver with regard to stress regulation (e.g., Spangler & Schieche, 1998). In addition, children and adults with insecure or disorganized attachment representations can also not effectively use social support from persons other than the main caregivers (Maunder & Hunter, 2001), since the internal working model is usually transmitted to all further close relationships, such as to teachers or therapists (Howes & Hamilton, 1992; Sroufe, Egeland, Carlson, & Collins, 2005). This is relevant because 60% to 90% of clinical samples or populations in schools for special education (Julius, 2001; van IJzendoorn & Bakermans-Kranenburg, 1996) show an insecure and/or disorganized attachment. In the typically developing population, secure attachment has been found in only 50% (Grossmann, Grossmann, Huber, & Wartner, 1981; German sample) to 60% (Ainsworth, Blehar, Waters, & Wall, 1978; U.S. sample) of children. In general, within the model of developmental psychopathology, secure attachment is seen as a protective factor (Werner & Smith, 1982), whereas insecure and disorganized attachment represents risk factors for socioemotional development (Strauss, Buchheim, & Kächele, 2002).

The regulation of stress and anxiety is a central feature not only of attachment theory but also of social support. Within social support theory, information support and instrumental support are distinguished from emotional support. Also, physical contact can be a means of expressing social support (Ditzen et al., 2007) and is most frequently found in close and trusting relationships, such as secure attachment, but less often in dyads with insecure attachment (Hazan & Zeifman, 1999). Most effective in reducing autonomic and endocrinological stress responses is a combination of emotional support with physical contact (Demakis & McAdams, 1994; Ditzen et al., 2007).

In comparison with seeking attachment and social support from other humans, there are three advantages of the use of HAI as a therapeutic aid. First, the internal working model of attachment to the main caregivers is generally not spontaneously transmitted to animals (Kurdek, 2008, 2009), and development of a secure attachment to the companion animal seems possible. Indeed, many children and adults turn to their companion animals for social support when stressed (McNicholas & Collis, 2006; Rost & Hartmann, 1994). Frequently, children communicate critical personal matters to their pets rather than to other humans (Kurdek, 2008, 2009; Parish-Plass, 2008). Relationships with companion animals can indeed meet the criteria for attachment according to Ainsworth (1991). Second, although among humans in Western societies physical contact may be restricted to close and trusting relationships, it is a natural concomitant of HAI. It is easily established even with unfamiliar but friendly animals in a first-time encounter and can contribute to stress reduction (Beetz, Julius, et al., 2012; Beetz et al., 2011). Finally, animals provide a good opportunity to display caregiving behavior. Caregiving is the behavior in the parent that is complementary to the child's attachment system and that already starts to develop during childhood (George & Solomon, 2008; Solomon & George, 1999). Providing successful care and protection, including social support, has been linked to feelings of satisfaction and joy but also to the reduction of stress in the caregiver in both humans and animals (Bales, Kramer, Lewis-Reese, & Carter, 2006; George & Solomon, 2008; Kenkel et al., 2012). In humans and animals, the OT system is involved in the behavioral expression of caregiving (Bales, Kim, Lewis-Reese, & Carter, 2004; Bales et al., 2011; Feldman et al., 2011; Kenkel et al., 2012). Thus, grooming and feeding an animal might be associated with positive effects similar to receiving social support from it (Dunbar, 2010)—indeed, it seems as if attachment and caregiving behaviors are difficult to distinguish in HAI (Julius et al., 2013).

Polyvagal Theory

Another theoretical construct that aids us in understanding social engagement, whether between humans or between humans and animals, is polyvagal theory (Porges, 1995, 2001, 2009; see also Chapter 4, this volume). The traditional view of the parasympathetic nervous system is of a vegetative system that promotes growth and homeostasis, and acts in opposition to the sympathetic nervous system's fight or flight response (Langley, 1921). The polyvagal theory recognizes that in mammals, the vagal nerve consists of two branches, a myelinated branch that innervates the heart and bronchi and an unmyelinated branch that regulates the facial muscles, which are so crucial for social communication (Ritz, Claussen, & Dahme, 2001). The evolutionary basis of this theory ties phylogenetic changes in these neural systems with the uniquely mammalian coupling between visceral states (safety vs. defensiveness) and the readiness for social engagement (Demaree, Robinson, Everhart, & Schmeichel, 2004; Demaree et al., 2006; Miskovic & Schmidt, 2010; Park, Van Bavel, Vasey, Egan, & Thayer, 2012; Thayer & Lane, 2000). Thus, there is central and peripheral cross-talk, mediated via the vagus nerve, which affects our ability to engage in social relationships. This evolutionary development may also be regulated by the OT system.

OT can regulate the autonomic nervous system, possibly contributing to interactions between social interactions and affective states (Grippo et al., 2012; Grippo, Trahanas, Zimmerman, Porges, & Carter, 2009; Norman, Cacioppo, Morris, Malarkey, et al., 2011). In prairie voles, peripheral OT administration decreased autonomic consequences of isolation, including decreasing heart rate and increasing heart-rate variability, a measure of parasympathetic cardiac control (Grippo et al., 2009, 2012). In humans, intranasal administration of OT also increases heart-rate variability; however, higher levels of loneliness were associated with lower levels of parasympathetic cardiac control (Norman, Cacioppo, Morris, Karelina, et al., 2011).

Thus, the polyvagal theory helps us understand the biological and evolutionary substrates on which OT can act to allow social engagement and the ability to engage in "immobility without fear" (see Chapter 4, this volume), both of which are critical components in social relationships, including human–animal relationships. Previous HAI studies that measured autonomic variables have often been limited to heart rate (reviewed in Beetz, Uvnäs-Moberg, et al., 2012), usually showing an effect of HAI on lowering heart rate. Research studies on the effects of HAI on the autonomic nervous system, including variables specific to parasympathetic regulation (e.g., heart-rate variability), represent a fertile area for future research.

Classical Conditioning and Distraction Effects

Although it seems likely that human–animal interaction is mediated through physiological and neural systems adapted for other types of social behavior (i.e., attachment to other humans), the possibility exists that other psychological phenomena may account for the observed health benefits. Ideally, research designs would include controls for these alternatives, which include classical conditioning and distraction effects. Classical conditioning implies that the human comes to associate the animal with relaxation, and it thus becomes a conditioned stimulus that can evoke physiological responses. Although this hypothesis makes theoretical sense, there are few empirical experiments on this topic, according to Virués-Ortega and Buela-Casal (2006). The idea of animals as distraction applies mainly to the context of AAI, in which the presence of the animal may serve to lessen the experience of anxiety or pain. Other, possibly simpler, distractions, including cartoons (Lee et al., 2012), music (Kleiber & Adamek, 2013), and clowns (Vagnoli, Caprilli, & Messeri, 2010), and serious (i.e., cognitively and behaviorally engaging) video games (Patel et al., 2006; Nilsson, Enskär, Hallqvist, & Kokinsky, 2013), may also be effective. More active/engaging pastimes appear to be more distracting (Nilsson et al., 2013) and are generally more effective than a different attachment object such as the parent (Scully, 2012). It is thus possible that the main buffering effects of a distraction will be dependent on what is particularly engaging for the patient in question. If the effects of animal presence are mainly as a distractor, there often may be simpler or cheaper alternatives for some patients.

THE ANIMAL'S SIDE IN HAI AND RELATIONSHIPS

Very few studies have included the animal side when investigating HAI. This is surprising when considering the large variation in individuals and species and their potentially strong influence on all variables regarding the human–animal dyad. Particularly, the more common companion animal species may share the basic behavioral systems of attachment and similar social tools and needs with humans (Julius et al., 2013). On this basis they can build real social relationships with humans and may profit in a similar way from HAI. For dogs, there are preliminary data indicating that the behavior toward the owner in an experimental test situation can be understood in terms of attachment (Topál et al., 2005; Topál, Miklósi, Csányi, & Dóka, 1998). In accordance with the observation that in adult-to-adult partner relationships, as well as dog–owner dyads, roles of social supporter and supportee are switched depending on the situation. Custance and Mayer (2012) found that dogs may show empathic-like concern and "consolation behavior" toward humans displaying distress. Also, studies have found positive effects of interactions with humans for dogs, such as an increase in OT levels and decrease in heart rate after being stroked (Handlin et al., 2011; Odendaal, 2000). This suggests that the same neurobiological mechanisms that underlie relationships between humans are linked to interactions and relationships between dogs and their owners. Overall, the data support the assumption that the human–animal relationship can be bidirectional, depending on the animal species involved.

Another aspect that must be considered is the well-being and health of the animal involved in HAI and in particular in AAI. It is plausible that only an animal that is healthy, feeling well, and not overly stressed will socialize well with humans and can thus promote the beneficial effects. Today, this aspect receives increasing attention in parallel to the development of professional quality standards of AAI and against the background of the idea of One Health, the interdisciplinary movement to achieve optimal health for humans and animals (and the environment). (See also Chapter 10 in this volume, which addresses animal health and welfare in some detail.)

IMPLICATIONS FOR RESEARCH METHODOLOGY IN HAI

As HAI studies become increasingly biological and mechanistic, researchers have searched for relatively simple, noninvasive measures of physiological (and ideally neural) activity that can be used with sensitive target populations such as children and companion animals. These have included salivary and urinary measures of hormones (cortisol, OT); autonomic measures (heart rate, heart-rate variability); and potentially, electrophysiological or neuroimaging measures (EEG, fMRI, PET).

Cortisol, as a potential biomarker for stress, is a prime candidate for examination of the health effects of HAI. It is also one of the best-studied hormones and easiest to collect noninvasively, not requiring any expensive equipment or even a freezer at the site of collection (see Chapter 9, this volume). This steroid hormone is a measurable output of activation of the hypothalamic–pituitary–adrenal (HPA) axis, a system that responds to challenging physiological and behavioral events by releasing glucocorticoids (including cortisol) to regulate glucose use by the body (for a recent review, see Hawkley, Cole, Capitanio, Norman, & Cacioppo, 2012). Although it may be easy to obtain and measure, many challenges of collection and interpretation remain (Hellhammer, Wüst, & Kudielka, 2009; Levine, Zagoory-Sharon, Feldman, Lewis, & Weller, 2007; see also Chapter 9, this volume).

Salivary cortisol is typically considered the gold standard for stress research (Hellhammer et al., 2009; Marques, Silverman, & Sternberg, 2010). Considerations in its collection may include the area of the mouth where it was collected, food eaten before or used to induce salivation during collection, medication effects, and subject compliance in home collections (Adam & Kumari, 2009; Granger, Hibel, Fortunato, & Kapelewski, 2009; Jessop & Turner-Cobb, 2008; see also Chapter 9, this volume). Just one collection of baseline cortisol is unlikely to adequately represent the functioning of the system—potential measures include the cortisol awakening response, reactivity to social or nonsocial stressors, the diurnal cortisol slope, and the area under the daytime cortisol curve (Adam & Kumari, 2009).

Cardiovascular measures, as representative of the autonomic nervous system, may also end up being valuable to our understanding of the physiological effects of HAI. Measures such as heart rate, and heart-rate variability, although potentially complicated to analyze (Chapleau & Sabharwal, 2011), are also easy to collect relatively noninvasively. These measures become important as ways to assess engagement of the parasympathetic nervous system and its relationship to our ability to engage socially.

The neural mechanisms involved in HAI research are one of the remaining great frontiers. At this time, there are few available studies examining either the neural responses of humans to animals or vice versa (although interest appears to be growing in this area; see Berns, Brooks, & Spivak, 2015; Cook, Spivak, & Berns, 2014; Stoeckel, Palley, Gollub, Niemi, & Evins, 2014). These basic studies could be accomplished perhaps more easily with humans, who would not need training or anesthesia to stay still in an MRI scanner. Imaging or even EEG studies would aid in determining whether neural substrates of the human–animal bond are more closely related to those displayed in other emotional bonds such as "love," pair-bonds, friendships, and mother–infant relationships.

CONCLUSION

Affiliation and attachment represent significant aspects of human–animal relationships and are closely related to the positive effects of HAI in connection with companion animal ownership and AAIs. Because of their close connection with physiology and psychological functions, affiliation and attachment could be valuable as theoretical background for HAI research as well as for AAI. Furthermore, the bond also points to the importance of both partners of the interspecies dyad. Future research might profit from a systems approach including the side of the animal, with regard to an explanation for the variation in findings and identifying optimal features of the animals involved in AAI.

REFERENCES

Adam, E. K., & Kumari, M. (2009). Assessing salivary cortisol in large-scale, epidemiological research. *Psychoneuroendocrinology*, 34, 1423–1436. http://dx.doi.org/10.1016/j.psyneuen.2009.06.011

Ainsworth, M. D. S. (1963). The development of infant–mother interaction among the Ganda. In B. M. Foss (Ed.), *Determinants of infant behavior* (pp. 67–104). New York, NY: Wiley.

Ainsworth, M. D. S. (1991). Attachment and other affectional bonds across the life cycle. In C. Parkes, J. Stevenson-Hinde, & P. Marris (Eds.), *Attachment across the life cycle* (pp. 33–51). New York, NY: Routledge.

Ainsworth, M. D. S., Bell, S. M., & Stayton, D. J. (1971). Individual differences in strange-situation behavior of one-year-olds. In H. R. Schaffer (Ed.), *The origins of human social relations* (pp. 17–52). New York, NY: Academic Press.

Ainsworth, M. D. S., Blehar, M. C., Waters, E., & Wall, S. (1978). *Patterns of attachment: A psychological study of the strange situation.* Hillsdale, NJ: Erlbaum.

Ainsworth, M. D. S., & Wittig, B. A. (1969). Attachment and the exploratory behavior of one-year-olds in a strange situation. In B. M. Foss (Ed.), *Determinants of infant behavior* (pp. 111–136). London, England: Methuen.

Allen, K. M., Blascovich, J., & Mendes, W. B. (2002). Cardiovascular reactivity and the presence of pets, friends, and spouses: The truth about cats and dogs. *Psychosomatic Medicine, 64*, 727–739.

Allen, K. M., Blascovich, J., Tomaka, J., & Kelsey, R. M. (1991). Presence of human friends and pet dogs as moderators of autonomic responses to stress in women. *Journal of Personality and Social Psychology, 61*, 582–589. http://dx.doi.org/10.1037/0022-3514.61.4.582

Bales, K. L., Boone, E., Epperson, P., Hoffman, G., & Carter, C. S. (2011). Are behavioral effects of early experience mediated by oxytocin? *Frontiers in Psychiatry, 2*, 24. http://dx.doi.org/10.3389/fpsyt.2011.00024

Bales, K. L., Kim, A. J., Lewis-Reese, A. D., & Carter, C. S. (2004). Both oxytocin and vasopressin may influence alloparental behavior in male prairie voles. *Hormones and Behavior, 45*, 354–361. http://dx.doi.org/10.1016/j.yhbeh.2004.01.004

Bales, K. L., Kramer, K. M., Lewis-Reese, A. D., & Carter, C. S. (2006). Effects of stress on parental care are sexually dimorphic in prairie voles. *Physiology & Behavior, 87*, 424–429. http://dx.doi.org/10.1016/j.physbeh.2005.11.002

Beetz, A., Julius, H., Turner, D., & Kotrschal, K. (2012). Effects of social support by a dog on stress modulation in male children with insecure attachment. *Frontiers in Psychology, 3*, 352. http://dx.doi.org/10.3389/fpsyg.2012.00352

Beetz, A., Kotrschal, K., Turner, D., Hediger, K., Uvnäs-Moberg, K., & Julius, H. (2011). The effect of a real dog, toy dog and friendly person on insecurely attached children during a stressful task: An exploratory study. *Anthrozoös, 24*, 349–368. http://dx.doi.org/10.2752/175303711X13159027359746

Beetz, A., Uvnäs-Moberg, H., Julius, K., & Kotrschal, K. (2012). Psychosocial and psychophysiological effects of human-animal interactions: The possible role of oxytocin. *Frontiers in Psychology, 3*. http://dx.doi.org/10.3389/fpsyg.2012.00234

Berns, G. S., Brooks, A. M., & Spivak, M. (2015). Scent of the familiar: An fMRI study of canine brain responses to familiar and unfamiliar human and dog odors. *Behavioural Processes, 110*, 37–46.

Bodnar, R. J., Nilaver, G., Wallace, M. M., Badillo-Martinez, D., & Zimmerman, E. A. (1984). Pain threshold changes in rats following central injection of beta-endorphin, met-enkephalin, vasopressin or oxytocin antisera. *International Journal of Neuroscience, 24*, 149–160. http://dx.doi.org/10.3109/00207458409089803

Bowlby, J. (1969). *Attachment and Loss: Vol. 1. Attachment.* New York, NY: Basic Books.

Bretherton, I., & Munholland, K. A. (2008). Internal working models in attachment relationships: Elaborating a central construct in attachment theory. In J. Cassidy & P. R. Shaver (Eds.), *Handbook of attachment: Theory, research and clinical applications* (2nd ed., pp. 134–152). New York, NY: Guilford Press.

Cardoso, C., Ellenbogen, M. A., Orlando, M. A., Bacon, S. L., & Joober, R. (2013). Intranasal oxytocin attenuates the cortisol response to physical stress: A dose-response study. *Psychoneuroendocrinology, 38*, 399–407. http://dx.doi.org/10.1016/j.psyneuen.2012.07.013

Chapleau, M. W., & Sabharwal, R. (2011). Methods of assessing vagus nerve activity and reflexes. *Heart Failure Reviews, 16*, 109–127. http://dx.doi.org/10.1007/s10741-010-9174-6

Cho, M. M., DeVries, A. C., Williams, J. R., & Carter, C. S. (1999). The effects of oxytocin and vasopressin on partner preferences in male and female prairie voles (*Microtus ochrogaster*). *Behavioral Neuroscience, 113*, 1071–1079. http://dx.doi.org/10.1037/0735-7044.113.5.1071

Cook, P. F., Spivak, M., & Berns, G. S. (2014). One pair of hands is not like another: Caudate BOLD response in dogs depends on signal source and canine temperament. *PeerJ, 2*, e596. http://dx.doi.org/10.7717/peerj.596

Custance, D., & Mayer, J. (2012). Empathic-like responding by domestic dogs (*Canis familiaris*) to distress in humans: An exploratory study. *Animal Cognition, 15*, 851–859. http://dx.doi.org/10.1007/s10071-012-0510-1

DeLoache, J. S., Pickard, M. B., & LoBue, V. (2011). How very young children think about animals. In P. McCardle, S. McCune, J. A. Griffin, & V. Maholmes (Eds.), *How animals affect us: Examining the influence of human-animal interaction on child development and human health* (pp. 85–99). Washington, DC: American Psychological Association. http://dx.doi.org/10.1037/12301-004

Demakis, G. J., & McAdams, D. P. (1994). Personality, social support and well-being among first year college students. *College Student Journal, 28*, 235–243.

Demaree, H. A., Robinson, J. L., Everhart, D. E., & Schmeichel, B. J. (2004). Resting RSA is associated with natural and self-regulated responses to negative emotional stimuli. *Brain and Cognition, 56*, 14–23. http://dx.doi.org/10.1016/j.bandc.2004.05.001

Demaree, H. A., Schmeichel, B. J., Robinson, J. L., Pu, J., Everhart, D. E., & Berntson, G. G. (2006). Up- and down-regulating facial disgust: Affective, vagal, sympathetic, and respiratory consequences. *Biological Psychology, 71*, 90–99. http://dx.doi.org/10.1016/j.biopsycho.2005.02.006

Ditzen, B., Neumann, I. D., Bodenmann, G., von Dawans, B., Turner, R. A., Ehlert, U., & Heinrichs, M. (2007). Effects of different kinds of couple interaction on cortisol and heart rate responses to stress in women. *Psychoneuroendocrinology*, *32*, 565–574. http://dx.doi.org/10.1016/j.psyneuen.2007.03.011

Dunbar, R. I. (2010). The social role of touch in humans and primates: Behavioural function and neurobiological mechanisms. *Neuroscience and Biobehavioral Reviews*, *34*, 260–268. http://dx.doi.org/10.1016/j.neubiorev.2008.07.001

Feldman, R., Gordon, I., & Zagoory-Sharon, O. (2011). Maternal and paternal plasma, salivary, and urinary oxytocin and parent-infant synchrony: Considering stress and affiliation components of human bonding. *Developmental Science*, *14*, 752–761. http://dx.doi.org/10.1111/j.1467-7687.2010.01021.x

George, C., & Solomon, J. (2008). The caregiving system: A behavioral systems approach to parenting. In J. Cassidy & P. Shaver (Eds.), *Handbook of attachment: Theory, research and clinical applications* (pp. 833–856). New York, NY: Guilford Press.

Granger, D. A., Hibel, L. C., Fortunato, C. K., & Kapelewski, C. H. (2009). Medication effects on salivary cortisol: Tactics and strategy to minimize impact in behavioral and developmental science. *Psychoneuroendocrinology*, *34*, 1437–1448. http://dx.doi.org/10.1016/j.psyneuen.2009.06.017

Grippo, A. J., Pournajafi-Nazarloo, H., Sanzenbacher, L., Trahanas, D. M., McNeal, N., Clarke, D. A., . . . Carter, S. C. (2012). Peripheral oxytocin administration buffers autonomic but not behavioral responses to environmental stressors in isolated prairie voles. *Stress*, *15*, 149–161.

Grippo, A. J., Trahanas, D. M., Zimmerman, R. R., II, Porges, S. W., & Carter, C. S. (2009). Oxytocin protects against negative behavioral and autonomic consequences of long-term social isolation. *Psychoneuroendocrinology*, *34*, 1542–1553. http://dx.doi.org/10.1016/j.psyneuen.2009.05.017

Grossmann, K., Grossmann, K., Huber, F., & Wartner, U. (1981). German children's behavior towards their mothers at 12 months and their fathers at 18 months in Ainsworth's Strange Situation. *International Journal of Behavioral Development*, *4*, 157–181. http://dx.doi.org/10.1177/016502548100400202

Guastella, A. J., & MacLeod, C. (2012). A critical review of the influence of oxytocin nasal spray on social cognition in humans: Evidence and future directions. *Hormones and Behavior*, *61*, 410–418. http://dx.doi.org/10.1016/j.yhbeh.2012.01.002

Handlin, L., Hydbring-Sandberg, E., Nilsson, A., Ejdebäck, M., Jansson, A., & Uvnäs-Moberg, K. (2011). Short-term interaction between dogs and their owners: Effects on oxytocin, cortisol, insulin and heart rate—an exploratory study. *Anthrozoös*, *24*, 301–315. http://dx.doi.org/10.2752/175303711X13045914865385

Hawkley, L. C., Cole, S. W., Capitanio, J. P., Norman, G. J., & Cacioppo, J. T. (2012). Effects of social isolation on glucocorticoid regulation in social mammals. *Hormones and Behavior*, *62*, 314–323. http://dx.doi.org/10.1016/j.yhbeh.2012.05.011

Hazan, C., & Zeifman, D. (1999). Pair bonds as attachments: Evaluating the evidence. In J. Cassidy & P. R. Shaver (Eds.), *Handbook of attachment: Theory, research and clinical application* (pp. 436–455). New York, NY: Guilford Press.

Heinrichs, M., Baumgartner, T., Kirschbaum, C., & Ehlert, U. (2003). Social support and oxytocin interact to suppress cortisol and subjective responses to psychosocial stress. *Biological Psychiatry, 54*, 1389–1398. http://dx.doi.org/10.1016/S0006-3223(03)00465-7

Hellhammer, D. H., Wüst, S., & Kudielka, B. M. (2009). Salivary cortisol as a biomarker in stress research. *Psychoneuroendocrinology, 34*, 163–171. http://dx.doi.org/10.1016/j.psyneuen.2008.10.026

Howes, C., & Hamilton, C. E. (1992). Children's relationships with child care teachers: Stability and concordance with parental attachments. *Child Development, 63*, 867–878. http://dx.doi.org/10.2307/1131239

Insel, T. R. (2010). The challenge of translation in social neuroscience: A review of oxytocin, vasopressin, and affiliative behavior. *Neuron, 65*, 768–779. http://dx.doi.org/10.1016/j.neuron.2010.03.005

Jessop, D. S., & Turner-Cobb, J. M. (2008). Measurement and meaning of salivary cortisol: A focus on health and disease in children. *Stress, 11*, 1–14. http://dx.doi.org/10.1080/10253890701365527

Julius, H. (2001). Die Bindungsorganisation von Kindern, die an Erziehungshilfe schulen unterrichtet werden [Attachment organization in children educated at educational support schools]. *Sonderpädagogik, 31*, 74–93.

Julius, H., Beetz, A., Kotrschal, K., Turner, D., & Uvnäs-Moberg, K. (2013). *Attachment to pets*. New York, NY: Hogrefe.

Kellert, S. R., & Wilson, E. O. (1993). *The biophilia hypothesis*. Washington, DC: Island Press.

Kenkel, W. M., Paredes, J., Yee, J. R., Pournajafi-Nazarloo, H., Bales, K. L., & Carter, C. S. (2012). Neuroendocrine and behavioural responses to exposure to an infant in male prairie voles. *Journal of Neuroendocrinology, 24*, 874–886. http://dx.doi.org/10.1111/j.1365-2826.2012.02301.x

Kleiber, C., & Adamek, M. S. (2013). Adolescents' perceptions of music therapy following spinal fusion surgery. *Journal of Clinical Nursing, 22*, 414–422. http://dx.doi.org/10.1111/j.1365-2702.2012.04248.x

Kurdek, L. A. (2008). Pet dogs as attachment figures. *Journal of Social and Personal Relationships, 25*, 247–266. http://dx.doi.org/10.1177/0265407507087958

Kurdek, L. A. (2009). Pet dogs as attachment figures for adult owners. *Journal of Family Psychology, 23*, 439–446. http://dx.doi.org/10.1037/a0014979

Langley, J. N. (1921). *The autonomic nervous system*. Cambridge, England: Hefter & Sons.

Lee, J., Lee, J., Lim, H., Son, J.-S., Lee, J.-R., Kim, D.-C., & Ko, S. (2012). Cartoon distraction alleviates anxiety in children during induction of anesthesia. *Anesthesia and Analgesia, 115*, 1168–1173. http://dx.doi.org/10.1213/ANE.0b013e31824fb469

Levine, A., Zagoory-Sharon, O., Feldman, R., Lewis, J. G., & Weller, A. (2007). Measuring cortisol in human psychobiological studies. *Physiology & Behavior, 90*, 43–53. http://dx.doi.org/10.1016/j.physbeh.2006.08.025

Marques, A. H., Silverman, M. N., & Sternberg, E. M. (2010). Evaluation of stress systems by applying noninvasive methodologies: Measurements of neuro-immune biomarkers in the sweat, heart rate variability and salivary cortisol. *Neuro-immunomodulation, 17*, 205–208. http://dx.doi.org/10.1159/000258725

Marvin, R. S., & Britner, P. A. (2008). Normative development: The ontogeny of attachment. In J. Cassidy & P. R. Shaver (Eds.), *Handbook of attachment: Theory, research and clinical applications* (2nd ed., pp. 269–294). New York, NY: Guilford Press.

Maunder, R. G., & Hunter, J. J. (2001). Attachment and psychosomatic medicine: Developmental contributions to stress and disease. *Psychosomatic Medicine, 63*, 556–567. http://dx.doi.org/10.1097/00006842-200107000-00006

McNicholas, J., & Collis, G. (2006). Animals as social supporters. Insights for understanding animal-assisted therapy. In A. Fine (Ed.), *A handbook on animal-assisted therapy* (pp. 49–71). San Diego, CA: Elsevier.

Miskovic, V., & Schmidt, L. A. (2010). Frontal brain electrical asymmetry and cardiac vagal tone predict biased attention to social threat. *International Journal of Psychophysiology, 75*, 332–338. http://dx.doi.org/10.1016/j.ijpsycho.2009.12.015

Nagasawa, M., Kikusui, T., Onaka, T., & Ohta, M. (2009). Dog's gaze at its owner increases owner's urinary oxytocin during social interaction. *Hormones and Behavior, 55*, 434–441. http://dx.doi.org/10.1016/j.yhbeh.2008.12.002

Neumann, I. D., & Landgraf, R. (2012). Balance of brain oxytocin and vasopressin: Implications for anxiety, depression, and social behaviors. *Trends in Neurosciences, 35*, 649–659. http://dx.doi.org/10.1016/j.tins.2012.08.004

Nilsson, S., Enskär, K., Hallqvist, C., & Kokinsky, E. (2013). Active and passive distraction in children undergoing wound dressings. *Journal of Pediatric Nursing, 28*, 158–166. http://dx.doi.org/10.1016/j.pedn.2012.06.003

Norman, G. J., Cacioppo, J. T., Morris, J. S., Karelina, K., Malarkey, W. B., Devries, A. C., & Berntson, G. G. (2011). Selective influences of oxytocin on the evaluative processing of social stimuli. *Journal of Psychopharmacology, 25*, 1313–1319. http://dx.doi.org/10.1177/0269881110367452

Norman, G. J., Cacioppo, J. T., Morris, J. S., Malarkey, W. B., Berntson, G. G., & Devries, A. C. (2011). Oxytocin increases autonomic cardiac control: Moderation by loneliness. *Biological Psychology, 86*, 174–180. http://dx.doi.org/10.1016/j.biopsycho.2010.11.006

Odendaal, J. S. (2000). Animal-assisted therapy—magic or medicine? *Journal of Psychosomatic Research, 49*, 275–280. http://dx.doi.org/10.1016/S0022-3999(00)00183-5

Odendaal, J. S., & Meintjes, R. A. (2003). Neurophysiological correlates of affiliative behaviour between humans and dogs. *Veterinary Journal, 165*, 296–301. http://dx.doi.org/10.1016/S1090-0233(02)00237-X

Parish-Plass, N. (2008). Animal-assisted therapy with children suffering from insecure attachment due to abuse and neglect: A method to lower the risk of intergenerational transmission of abuse? *Clinical Child Psychology and Psychiatry, 13,* 7–30. http://dx.doi.org/10.1177/1359104507086338

Park, G., Van Bavel, J. J., Vasey, M. W., Egan, E. J., & Thayer, J. F. (2012). From the heart to the mind's eye: Cardiac vagal tone is related to visual perception of fearful faces at high spatial frequency. *Biological Psychology, 90,* 171–178. http://dx.doi.org/10.1016/j.biopsycho.2012.02.012

Patel, A., Schieble, T., Davidson, M., Tran, M. C., Schoenberg, C., Delphin, E., & Bennett, H. (2006). Distraction with a hand-held video game reduces pediatric preoperative anxiety. *Paediatric Anaesthesia, 16,* 1019–1027. http://dx.doi.org/10.1111/j.1460-9592.2006.01914.x

Pedersen, C. A., & Prange, A. J., Jr. (1979). Induction of maternal behavior in virgin rats after intracerebroventricular administration of oxytocin. *Proceedings of the National Academy of Sciences, USA, 76,* 6661–6665. http://dx.doi.org/10.1073/pnas.76.12.6661

Petersson, M., & Uvnäs-Moberg, K. (2007). Effects of an acute stressor on blood pressure and heart rate in rats pretreated with intracerebroventricular oxytocin injections. *Psychoneuroendocrinology, 32,* 959–965. http://dx.doi.org/10.1016/j.psyneuen.2007.06.015

Petersson, M., & Uvnäs-Moberg, K. (2008). Postnatal oxytocin treatment of spontaneously hypertensive male rats decreases blood pressure and body weight in adulthood. *Neuroscience Letters, 440,* 166–169. http://dx.doi.org/10.1016/j.neulet.2008.05.091

Porges, S. W. (1995). Orienting in a defensive world: Mammalian modifications of our evolutionary heritage. A polyvagal theory. *Psychophysiology, 32,* 301–318. http://dx.doi.org/10.1111/j.1469-8986.1995.tb01213.x

Porges, S. W. (2001). The polyvagal theory: Phylogenetic substrates of a social nervous system. *International Journal of Psychophysiology, 42,* 123–146. http://dx.doi.org/10.1016/S0167-8760(01)00162-3

Porges, S. W. (2009). The polyvagal theory: New insights into adaptive reactions of the autonomic nervous system. *Cleveland Clinic Journal of Medicine, 76*(Suppl. 2), S86–S90. http://dx.doi.org/10.3949/ccjm.76.s2.17

Ritz, T., Claussen, C., & Dahme, B. (2001). Experimentally induced emotions, facial muscle activity, and respiratory resistance in asthmatic and non-asthmatic individuals. *British Journal of Medical Psychology, 74,* 167–182. http://dx.doi.org/10.1348/000711201160894

Ross, H. E., Cole, C. D., Smith, Y., Neumann, I. D., Landgraf, R., Murphy, A. Z., & Young, L. J. (2009). Characterization of the oxytocin system regulating affiliative behavior in female prairie voles. *Neuroscience, 162,* 892–903.

Rost, D. H., & Hartmann, A. (1994). Children and their pets. *Anthrozoös, 7,* 242–254. http://dx.doi.org/10.2752/089279394787001709

Scully, S. M. (2012). Parental presence during pediatric anesthesia induction. *AORN Journal, 96,* 26–33.

Seltzer, L. J., Ziegler, T. E., & Pollak, S. D. (2010). Social vocalizations can release oxytocin in humans. *Proceedings of the Royal Society B: Biological Sciences, 277,* 2661–2666. http://dx.doi.org/10.1098/rspb.2010.0567

Serpell, J. A. (1986). *In the company of animals.* Oxford, England: Blackwell.

Solomon, J., & George, C. (1999). *Attachment disorganization.* New York, NY: Guilford Press.

Spangler, G., & Schieche, M. (1998). Emotional and adrenocortical responses of infants to the strange situation: The differential function of emotional expression. *International Journal of Behavioral Development, 22,* 681–706. http://dx.doi.org/10.1080/016502598384126

Sroufe, L. A., Egeland, B., Carlson, E. A., & Collins, W. A. (2005). *The development of the person.* New York, NY: Guilford Press.

Stoeckel, L. E., Palley, L. S., Gollub, R. L., Niemi, S. M., & Evins, A. E. (2014). Patterns of brain activation when mothers view their own child and dog: An fMRI study. *PLoS ONE, 9,* e107205. Advance online publication. http://dx.doi.org/10.1371/journal.pone.0107205

Strauss, B., Buchheim, A., & Kächele, H. (2002). *Klinische Bindungsforschung. Theorien, Methoden, Ergebnisse* [Clinical attachment research. Theories, methods, results]. Stuttgart, Germany: Schattauer.

Thayer, J. F., & Lane, R. D. (2000). A model of neurovisceral integration in emotion regulation and dysregulation. *Journal of Affective Disorders, 61,* 201–216. http://dx.doi.org/10.1016/S0165-0327(00)00338-4

Topál, J., Gácsi, M., Miklósi, Á., Virányi, Z., Kubinyi, E., & Csányi, V. (2005). Attachment to humans: A comparative study on hand-reared wolves and differently socialized dog puppies. *Animal Behaviour, 70,* 1367–1375. http://dx.doi.org/10.1016/j.anbehav.2005.03.025

Topál, J., Miklósi, Á., Csányi, V., & Dóka, A. (1998). Attachment behavior in dogs (*Canis familiaris*): A new application of Ainsworth's (1969) Strange Situation Test. *Journal of Comparative Psychology, 112,* 219–229. http://dx.doi.org/10.1037/0735-7036.112.3.219

Uvnäs-Moberg, K. (2003). *The oxytocin factor: Tapping the hormone of calm, love, and healing.* Cambridge, MA: Da Capo Press.

Vagnoli, L., Caprilli, S., & Messeri, A. (2010). Parental presence, clowns or sedative premedication to treat preoperative anxiety in children: What could be the most promising option? *Paediatric Anaesthesia, 20,* 937–943. http://dx.doi.org/10.1111/j.1460-9592.2010.03403.x

van IJzendoorn, M. H., & Bakermans-Kranenburg, M. J. (1996). Attachment representations in mothers, fathers, adolescents, and clinical groups: A meta-analytic search for normative data. *Journal of Consulting and Clinical Psychology, 64,* 8–21. http://dx.doi.org/10.1037/0022-006X.64.1.8

van IJzendoorn, M. H., & Bakermans-Kranenburg, M. J. (2012). A sniff of trust: Meta-analysis of the effects of intranasal oxytocin administration on face recognition, trust to in-group, and trust to out-group. *Psychoneuroendocrinology*, *37*, 438–443. http://dx.doi.org/10.1016/j.psyneuen.2011.07.008

Virués-Ortega, J., & Buela-Casal, G. (2006). Psychophysiological effects of human-animal interaction: Theoretical issues and long-term interaction effects. *Journal of Nervous and Mental Disease*, *194*, 52–57. http://dx.doi.org/10.1097/01.nmd.0000195354.03653.63

Werner, E., & Smith, R. (1982). *Vulnerable but invincible: A study of resilient children*. New York, NY: McGraw-Hill.

Williams, J. R., Insel, T. R., Harbaugh, C. R., & Carter, C. S. (1994). Oxytocin administered centrally facilitates formation of a partner preference in female prairie voles (*Microtus ochrogaster*). *Journal of Neuroendocrinology*, *6*, 247–250. http://dx.doi.org/10.1111/j.1365-2826.1994.tb00579.x

Wilson, E. O. (1984). *Biophilia*. Cambridge, MA: Harvard University Press.

Yang, J., Li, P., Liang, J. Y., Pan, Y. J., Yan, X. Q., Yan, F. L., . . . Wang, D. X. (2011). Oxytocin in the periaqueductal grey regulates nociception in the rat. *Regulatory Peptides*, *169*, 39–42. http://dx.doi.org/10.1016/j.regpep.2011.04.007

Zingg, H. H. (2002). Oxytocin. In D. Pfaff, A. P. Arnold, A. M. Etgen, S. E. Fahrbach, & R. T. Rubin (Eds.), *Hormones, brain and behavior* (pp. 779–802). New York, NY: Academic Press. http://dx.doi.org/10.1016/B978-012532104-4/50059-7

6

THE SOCIAL REGULATION OF NEURAL THREAT RESPONDING

CASEY BROWN AND JAMES A. COAN

Social support improves psychological and physiological health, and social isolation increases mortality at rates rivaling well-established risk factors such as cigarette smoking, high blood pressure, high blood lipid concentrations, obesity, and lack of physical activity (House, Landis, & Umberson, 1988; Uchino, Cacioppo, & Kiecolt-Glaser, 1996). Although these associations are well established, the mechanisms through which social support exerts its effects remain unclear. This chapter outlines our research on the social regulation of neural threat responding, emphasizing how reliable social support economizes cognitive, affective, and behavioral effort. Specifically, we present *social baseline theory*, which posits that humans' default or baseline mode of affect regulation is via social proximity and that the human ecology is acutely social. Indeed, when identifying the dominant human habitat, specific terrestrial features seem unlikely—humans live just about everywhere on earth, subsisting on a great variety of diets and adapting themselves to a

http://dx.doi.org/10.1037/14856-008
The Social Neuroscience of Human–Animal Interaction, L. S. Freund, S. McCune, L. Esposito, N. R. Gee, and P. McCardle (Editors)

diversity of conditions. If humans have created a social ecology—an ecology based on a rich collection of social behaviors and capacities—it follows that other animals could adapt themselves to that ecology. We conjecture that household pets, with a specific emphasis on domesticated dogs, have indeed adapted to the human social ecology and that humans and domesticated pets form veridical social relationships with one another, to great mutual benefit. Companion animals may indeed fulfill the roll of a social support provider— even a relational partner—with consequences for the regulation of neurobiological mechanisms supporting the brain's threat response and, by extension, for the many consequences of that regulation for health and well-being, both human and animal. In this chapter, we first ground our perception in contemporary theories related to behavioral ecology and perception.

ECONOMY OF ACTION AND THE ECOLOGY OF PERCEPTION

Living organisms, to accomplish the evolutionary imperatives of survival and reproduction, must take in more energy than they consume. Thus, according to the *economy of action* principle (Proffitt, 2006), natural selection should favor adaptations that promote the conservation of energy. Like other animals, humans abide by this constraint and conserve their energy whenever possible.

According to the ecological view of perception, we perceive objects in terms of the actions they afford our bodies (Gibson, 1986). More recent research has suggested that perceptions of affordances also reflect our available resources (Proffitt, 2006). An association between human perception and factors that affect physical energy expenditure is likely an adaptation that promotes economy of action (Proffitt, 2006). Perception of the environment is malleable and affected by the energy required to act on the environment (Proffitt, 2006). Thus, perception is constrained by one's bioenergetic state (Schnall, Zadra, & Proffitt, 2010). Metabolic energy readily accessible to perform a given action fluctuates from moment to moment, due to fatigue, cognitive depletion, how recently we consumed calories, or other unmeasured individual differences, and they impact how we perceive the demands of the physical environment (Proffitt, Bhalla, Gossweiler, & Midgett, 1995). For example, levels of fatigue, depressed mood, stress, and circulating blood glucose can affect how steep hills appear (Schnall et al., 2010; Proffitt et al., 1995). When an individual's resources are taxed, as, for example, when wearing a heavy backpack, he or she will actually perceive a hill to be steeper (Bhalla & Proffitt, 1999) than when feeling rested and energetic. This shift in perception affects the likelihood of walking up that hill, a decision that reflects how available resources will be budgeted in light of expected returns.

Put simply, with fewer resources, more motivation is needed, and this need for increased motivation manifests as a hill that seems steeper.

EMOTION, SELF-REGULATION, AND RESOURCE BUDGETING

Our perception is also influenced by emotion. It has been suggested, for example, that emotions provide embodied information about the energy costs associated with potential actions (Zadra & Clore, 2011). Indeed, emotions may provide useful information about resources without requiring the use of higher level cognitions to determine consequences, making emotional influences on perception an efficient means of decision making (cf. Phelps & LeDoux, 2005; Zadra & Clore, 2011). That is, emotion may influence perception in a way that conserves resources, promoting economy of action while both decreasing the risk and increasing the chances of positive outcomes.

For adaptive human functioning, it is imperative that humans respond emotionally to salient environmental cues. Despite the adaptive value of emotions, humans live in a world where intense emotional reactions are not always appropriate. It is necessary to control and regulate emotional reactions in a complex social environment to behave adaptively and foster social connections. Human emotions are often driven by thoughts, interpretations, and imaginations, so that habitual or automatic means of controlling emotions are critical (Phelps & LeDoux, 2005). Luckily, our complex brains allow for executive functions like planning, inhibition, and the control of attention, all of which allow us to regulate emotion feelings and behaviors. But the self-regulation of emotion often involves effortful control, such as overriding impulses or manipulating one's thoughts and behaviors through behavioral suppression or cognitive reappraisal. These processes heavily use the prefrontal cortex area of the brain (Gross & Thompson, 2007; Wager, Davidson, Hughes, Lindquist, & Ochsner, 2008). Although effective self-regulation may relate to better health, social relationships, and increased life satisfaction (Haga, Kraft, & Corby, 2009; Lopes, Salovey, Côté, Beers, & Petty, 2005; Smyth & Arigo, 2009), it can also be costly. Indeed, it has been suggested that effortful self-control and regulation are only sustainable for short periods of time and are subject to *depletion*—a steady decrease in ability coupled with a steady increase in subjective exhaustion (Baumeister & Vohs, 2007).

Much research on self-regulation has focused on how cognitive effort on one task affects performance on subsequent self-regulatory tasks (Gailliot & Baumeister, 2007). Findings indicate that self-control is a limited resource, like a muscle that fatigues after exertion (Baumeister, Bratslavsky, Muraven, & Tice, 1998). Research has indicated that the depletion of self-control is dependent on glucose metabolism, a critical process in bioenergetics. Effortful cognitive tasks that require self-regulatory resources can cause measurable drops in blood glucose, but performance improves after the consumption of glucose (Gailliot et al., 2007).

Others have argued that glucose levels do not place literal bioenergetic constraints on self-regulation so much as serve to guide decision making based on predicted future availability of bodily and environmental sources of glucose (Kurzban, 2010). Intriguingly, a glucose mouth rinse—where glucose is detected but not actually absorbed—can lead to better self-control after depletion (Hagger & Chatzisarantis, 2013), suggesting that the brain uses information about the body's resources to make budgeting decisions about use of metabolic resources. The mouth rinse is a signal of available resources and allows us to anticipate future energy gains. Although much research supports the idea that self-regulation is depleting, more recent evidence suggests that whether depletion takes place depends on an individual's belief that willpower is a limited resource (Job, Dweck, & Walton, 2010). Taken together, findings suggest that it is not only *actual* resources that matter; rather, our perceptions and actions are rooted in part in predicted availability of resources, both within and outside of our bodies. Humans budget resources on the basis of predictions stemming from environmental signals indicating the availability or lack of resources; this anticipatory budgeting alters our perceptions of affordances in the environment that promote economy of action.

PERCEPTION OF SOCIAL RELATIONSHIPS AS RESOURCES

Similarly to fatigue or physical burdens, social resources may alter our perceptions, adaptively promoting economy of action. A social bond may signal a resource that can extend into the future, providing energy benefits. Indeed, similar to blood glucose and physical resources, social resources can also change the way one views a hill, which may be estimated as less steep when the climber is accompanied by a friend (Schnall, Harber, Stefanucci, & Proffitt, 2008). Similar effects were obtained even when the friend was not physically present—just the thought of a supportive friend could decrease the slant of a hill by a small degree. And the better the friend, the less steep the hill (Schnall et al., 2008). We interpret these effects to mean that social support signals the presence of additional resources and changes our perception of environmental affordances accordingly.

Elsewhere, we have proposed that social relationships may serve a variety of energy-saving functions and that humans use social relationships to decrease metabolic costs. The related processes of *risk distribution* (distribution of risk across members of a group to minimize environmental threats while maximizing energy outcomes) and *load sharing* (interdependence and cooperative behaviors) are hypothesized to produce energy-saving benefits of social contact (see Beckes & Coan, 2011, for a review). Humans benefit especially from load sharing in social relationships because unlike most (possibly all) other animals, we are capable of shared intentionality, "collaborative interactions in which participants share psychological states with one another" (Tomasello & Carpenter, 2007, p. 121). Human children, unlike other primates, are often concerned with providing helpful information to others and forming shared attention, goals, and intentions (Tomasello & Carpenter, 2007). In our view, shared intentionality allows for cooperative behaviors that contribute powerfully to the management and optimization of metabolic resources in humans.

SOCIAL AFFECT REGULATION

Humans may also confer energy benefits from social relationships when it comes to regulating responses to threatening, stressful, and emotional stimuli. Indeed, social support may protect against the negative effects of stress by decreasing the perceived demands associated with stressful or threatening environments, in much the same way that social support can decrease the perceived slant of a hill (Brown, Oudekerke, Szwedo, & Allen, 2013; Coan, Schaefer, & Davidson, 2006).

As we have discussed, self-regulatory abilities are limited by depletion constraints. Self-regulation, mediated as it is through the prefrontal cortex (PFC), often functions like a muscle that can grow exhausted from overuse. This suggests that the PFC is costly and thus a target of conservation pressures. The PFC, known for its role in flexible but inefficient (e.g., slow, effortful) processes related to inhibitory control and executive functioning, is a relative luxury in evolutionary terms. Indeed, its work may be burdensome in situations in which action is immediately required. Thus, we suspect that the brain tends toward limited use of the PFC when exhausted, malnourished, or highly stressed; when action is immediately required; or indeed when the many processes supported by prefrontal function have been active already for long periods of time. This is all to save resources for purposes intimately tied to survival. Further, research has suggested that when a threatening stimulus becomes more imminent, neural activity shifts from the PFC to midbrain regions associated with reflexive behaviors (Mobbs et al., 2007). Research in sports neuroscience consistently reveals a shift in blood away from prefrontal regions during exercise ("induced hypofrontality"; Dietrich & Sparling, 2004). Thus, the work of the PFC may actually interfere with other motor, metabolic, and perceptual systems during both acute stress and high levels of physical activity. It is important to note that this does not suggest that the work of the PFC may not be valuable under those circumstances—only that it is relatively inefficient within a single individual.

We think that humans may be able to mitigate the inefficiency of prefrontal work by in effect outsourcing prefrontal load to supportive social networks. Our work investigating how the presence of social support affects the neural response to a threatening stimulus demonstrates that social support can decrease the impact of a stressor without prefrontally mediated self-regulatory effort (Coan et al., 2006). Specifically, happily married wives were placed in a functional magnetic resonance imaging scanner and threatened with a mild electric shock. Threat regions of the brain were less active when holding a partner's than a stranger's hand when alone. Moreover, the relationship quality predicted the extent to which neural threat activity decreased during handholding such that the better the quality of the relationship, the less threatening the shock.

The general pattern observed had been hypothesized, but one surprising result was that none of these regulatory effects were mediated through any part of the PFC. Indeed, it is often assumed that social contact should attenuate threat responding by engaging activity of frontal regions such as the ventromedial PFC, which in turn down-regulates activity in threat-related brain regions (Eisenberger et al., 2011). Many studies suggest that when emotion regulation is a self-regulatory task, the PFC becomes more active in inverse proportion to many of the same threat-responsive regions we observed in our handholding study. But we have found no evidence for the top-down prefrontal regulation of affect during the partner-handholding condition. Instead, the PFC was itself quieted by supportive handholding. Thus, supportive handholding seemed to achieve the same or similar regulatory ends without incurring any prefrontal cost. Although the specific neural mechanisms of social affect regulation remain a mystery, social support is likely an efficient and metabolically cost-effective means of regulating affect compared with self-regulation. If so, there should be both phylogenetic and ontogenetic pressures to forms social groups.

SOCIAL BASELINE THEORY

Most definitions, measures, and theories related to emotion regulation focus on the individual. So, taking the example of the handholding study, one might assume the alone condition to be the baseline state of threat responding and that adding a relational partner to the mix would constitute an experimental condition designed to enhance top-down self-regulatory abilities, perhaps through increased prefrontal activation. But there may be an entirely different explanation: If being surrounded by others in a social environment is the norm, then social proximity—even handholding—may be closer to the human baseline. It is probably true that most human emotion regulation is socially situated and occurs in social situations and contexts (Campos, Walle, Dahl, & Main, 2011; Gross, Richards, & John, 2006). According to social baseline theory, to be alone, especially in a threatening context, is to be far from our baseline state.

By contrast, when in our established, predictable social environments, or with those we are close to, less emotion activation and self-regulation is necessary because we are closer to our baseline condition. Indeed, the human brain may be designed to assume that it is embedded in a social network, and cues that social resources are distant may make the environment seem more difficult and metabolically demanding, much like donning a heavy backpack. Thus, social isolation leads to the perception that the environment presents more problems. Because human brains may assume social proximity and closeness to social resources, and because being alone or distanced from social resources may lead to extra energy expenditure via the use of the PFC, social contact may be an energy-efficient method of affect regulation. Thus, many adaptations that promote social interaction and proximity may have arisen in part because of the energy benefits that social relationships and cooperative behaviors confer to individuals. Additionally, social emotions (rejection, contempt, sympathy, or admiration) provide us easily accessible information about our social resources and may be embodied representations of the status of those resources allowing more accurate budgeting.

HUMAN SOCIAL ECOLOGY

An animal is a reflection of its ecology or a model of its terrestrial environment, with adaptations specially suited for the environment or ecology it evolved in. Humans, however, are not specifically adapted to any one unique terrestrial environment and can exist and thrive in almost every ecosystem on the Earth's surface—from tundra to taiga—in groups of other humans, suggesting that humans' dominant ecology is a richly social one (Berscheid, 2003). Recent evolutionary theory suggests that culture has played an active role in the evolution of human genes (Richerson, Boyd, & Henrich, 2010), and the idea of gene–culture coevolution further supports this idea. Social relationships, cooperative behavior, and interdependence have benefitted human energy use and allocation over evolutionary time such that humans could evolve to fit a social ecology, as opposed to a specific physical ecology. Thus, human adaptations for joint attention, shared goals, and interdependence are a reflection of the human's adaptation to an ecology defined by social resources. Additionally, conventional wisdom assumes that brains evolved to process and deal with ecological problem-solving tasks, and social brain theory (Dunbar, 1998) suggests that the large size of primate brains is reflective of human evolution to process and navigate complex social systems, indicating that complex social systems are the ecology to which we are adapted.

According to ecological theories of optimization, an animal's energy expenditure is optimized in proportion to its proximity to the environment to which it has adapted. Following from social baseline theory, humans should function optimally when in rich social environments. This hypothesis is supported by findings that various measures of health and well-being are improved by the presence of social relationships and rich social networks (e.g., Beals, Peplau, & Gable, 2009; Cohen & Janicki-Deverts, 2009; Gallagher & Vella-Brodrick, 2008; Holt-Lunstad, Smith, & Layton, 2010). This rich social ecology allows culture to flourish; through cultural processes, humans modify themselves and their environments so as to create an "artificial" physical ecological niche. Thus, the ecological pressures on humans are largely social in nature.

Especially given that humans modify their physical ecologies extensively to suit their social needs, other animals and organisms should be similarly capable of taking advantage of the human social ecology. Our most obvious example of this may be the domesticated dog.

THE DOMESTICATION OF DOGS

Animals have tapped into the human social ecology and evolved within it. In particular, the domestic dog has flourished in this ecological niche. According to the Humane Society of the United States (HSUS; 2011), approximately 78.2 million dogs are owned as pets, and 39% of U.S. households own at least one dog. Although there is debate regarding the exact mechanisms used by dogs for social learning (Mersmann, Tomasello, Call, Kaminski, & Taborsky, 2011), domestic dogs are clearly skilled in using communicative cues provided by humans (Bräuer, Kaminski, Riedel, Call, & Tomasello, 2006; Hare, Brown, Williamson, & Tomasello, 2002; Miklósi et al., 2004; Pongrácz et al., 2001; Topál, Byrne, Miklósi, & Csányi, 2006). They have likely gained this ability through a complex evolutionary process. Dogs and humans have coexisted within the same environment for at least 14,000 years (Clutton-Brock, 1995). Early on, dogs were likely provided with discarded food from humans as a means of garbage disposal, making the understanding of human gestures indicating the availability of food critical (Reid, 2009).

Some believe that domestication caused dogs' sensitivity to human gestures, with humans and dogs experiencing convergent evolution of social cognitive skills (Hare & Tomasello, 2005). The "domestication hypothesis" proposes that over the course of domestication, dogs evolved to maintain sensitivity to human communication that their nondomesticated counterparts (i.e., wolves) lack (Udell, Dorey, & Wynne, 2010). This has been brought into question, however, and a more likely hypothesis (the "two-stage hypothesis") has been asserted: That is, for an animal to be sensitive to human gestures and communication, it must first accept humans as social companions and be conditioned to follow human directives (Udell et al., 2010). In this context, dogs may have evolved a special preparedness to respond to social cues (Reid, 2009; Udell et al., 2010). This argument is supported with studies showing that when wolves are socialized with humans during a sensitive developmental period, they can actually outperform dogs under the same conditions in tests of social intelligence (Udell, Dorey, & Wynne, 2008). Dogs may have developed a predisposition to attend to the gestures of humans to quickly learn the association between human action—perhaps even human social gestures—and the availability of food (Reid, 2009).

Clear characteristics distinguish the varied modern domesticated dogs from the wolves they transformed from, including reduced aggression and improved social cognitive abilities, all of which suggest that behavioral (and probably morphological) characteristics desirable to humans were among the main targets of domestication (Coppinger & Coppinger, 2001; Ellegren, 2005). Axelsson and colleagues (2013) identified 36 genomic regions representative of targets for selection over the course of dog domestication, many of which contain genes relevant to brain function and nervous system development that may underlie the behavioral differences between dogs and wolves. Beyond these genes related to behavioral differences, Axelsson and colleagues also saw signals of selection in 10 genes that play key roles in starch digestion and fat metabolism, providing support for adaptations in modern dogs to thrive on starch-rich diets provided by humans (Axelsson et al., 2013). Genetic differences between wolves and modern dogs demonstrate the adaptations of dogs to the human social ecology. Dogs evolved to live and breed in human societies, and that human environment and social setting is now the domesticated dog's own ecological niche (Miklósi, Topál, & Csányi, 2004).

DOGS AND THEIR HUMAN COMPANIONS

The welfare of an animal is much determined by its life experience, which we humans cannot know. However, we are able to determine such components of a dog's life as nutrition, health, comfort, and level of apparent suffering. These physical components have been studied, and the diet and nutrition of pet dogs are probably better now than ever before. Most pet owners purchase dog food of higher nutritional value than dogs could achieve in the wild.

Although owners' decisions can interfere with the life and welfare of their pets (a negative example is tail docking; Noonan, Rand, Blackshaw, & Priest, 1996), veterinary surgeries have developed to allow safe and relatively painless treatments for a variety of painful conditions in dogs (Stafford, 2007). Most major infections, diseases, and parasites can be ameliorated, and dog owners spend an average of $248 on veterinary (vaccine and wellness) visits annually (HSUS, 2011). No doubt, human groups dedicated to the well-being of dogs are working to promote even stronger welfare for the minority of domestic dogs that receive little or no care.

In addition to nutrition and health, dogs also may develop valuable mental abilities. Genetic modification over time can be seen clearly in domestic species, and domesticated dogs may have developed the capacity for skills such as herding or companionship as adaptations to the human social ecology (Cooper et al., 2003). Social learning by dogs can reduce the costs involved in the acquisition of other resources or skills. Some have argued that dogs have experienced declines in problem solving compared with wolves, due to humans decreasing the selection on the basis of cognitive demands (Frank & Frank, 1985; Frank, Frank, Hasselbach, & Littleton, 1989). However, more recent research has suggested that dogs demonstrate well-developed cognitive abilities, and decrements on tasks compared with wolves are mainly due to the sensitivity a typical dog shows toward its owner—that is, dogs have a tendency to behave as socially dependent (Topál, Miklósi, & Csányi, 1997). Indeed, a fascinating observation is that dog performance can improve in the presence of a human. For example—and in a way that we feel nontrivially echoes the mitigation of human stress responding by proximity to other humans—dogs are better at solving novel problems in the presence of their owners, compared with the presence of strangers or when alone, regardless of their owners' knowledge of the task (Topál, Miklósi, Csányi, & Dóka, 1998). Indeed, it is proposed that dogs have developed an ability to act with humans as "social units" and are predisposed to form attachment bonds with humans (Topál, Miklósi, & Csányi, 1997), even to the extent that dogs realize physiological and endocrinological benefits from human contact (Coppola, Grandin, & Enns, 2006; Hennessy, Williams, Miller, Douglas, & Voith, 1998; Lynch & McCarthy, 1969).

HUMANS AND THEIR DOG COMPANIONS

Although dogs receive many benefits from being immersed in the human social ecology, humans also benefit substantially from the human–dog bond. There is a long history of association, for example, between humans and domesticated dogs *as companions* (as opposed to animals who are merely and more or less only useful for some purpose). Within the United States, 62% of households own a pet (American Pet Products Association, 2011). Pets can provide humans with obvious benefits—protector (e.g., scaring away burglars or reducing vermin), herd manager, hunting partner—but there is more support for the idea that dog ownership affects human health and well-being through the relationship that forms between dog and owner (McNicholas et al., 2005). The emotional bond between owner and pet can be as subjectively intense as many human relationships and may confer many of the same psychological benefits of companionship (Siegel, 1990). Indeed, many pets are regarded as full members of the family (HSUS, 2011).

Social support is critical for the psychological and physical well-being of humans, and pets—perhaps especially dogs—can contribute to social support via the same or similar relationship processes that facilitate social affect regulation. Not only are pets social catalysts leading to greater social contact between people (McNicholas & Collis, 2000), but pets themselves may be able to provide social support and companionship. Proximity to a trusted relational partner can, for humans especially, attenuate the neural threat response (Coan et al., 2006). It is likely, though untested at present, that the presence of a companion animal would similarly decrease neural threat responding, as evidenced by the presence of an animal having an effect on children's resting blood pressures and heart rates (Friedmann, Katcher, Thomas, Lynch, & Messent, 1983). Affiliation with animals may take advantage of the underlying neurobiological mechanisms that support social affect regulation between interacting humans. And a similar process may be true for pets—the better the perceived relationship with the human companion, the more benefit the pet, too, receives.

Indeed, social surrogacy and embodied cognition theories propose that cognitive associations with many kinds of nonhuman stimuli can be emotionally charged. People may seek out even nonsocial targets to fulfill social needs and avoid loneliness (Derrick, Gabriel, & Hugenberg, 2009). Moreover, perceptions of interpersonal intimacy can be associated with even relatively abstract perceptions, such as temperature—the physical warmth associated with something as mundane as holding a hot cup of coffee can increase feelings of interpersonal warmth without the person's awareness (Williams & Bargh, 2008). People's interactions with their pets often involve warm physical contact, and dogs are a source of physical contact and comfort (Jennings, 1997), much like a positive conversation with a friend. According to the Associated Press (2010), 25% of pet owners who are married or cohabiting report that their pet is "a better listener than their spouses." Animal companionship, like human companionship, may improve human health and well-being, preventing emotional distress through the social regulation of emotion without the costly involvement of the PFC.

The effects of pet companionship may be particularly pronounced for individuals who otherwise lack rich and supportive social networks. Lonely people are more likely to anthropomorphize their pets, both at home (Epley, Akalis, Waytz, & Cacioppo, 2008; Epley, Waytz, & Cacioppo, 2007) and when loneliness is induced in a laboratory setting (Epley et al., 2008). Pets may not only decrease feelings of loneliness but also offset feelings of rejection as effectively as a best friend (McConnell, Brown, Shoda, Stayton, & Martin, 2011). Dogs as social catalysts may be particularly effective in elderly or disabled individuals more at risk of social isolation and loneliness (Lane, McNicholas, & Collis, 1998; McNicholas et al., 2005).

All pet owners do not benefit equally from pet ownership; this has been shown for dog owners (McConnell et al., 2011). Some studies suggest that pet ownership confers *no* health benefits and may even be associated with poorer mental health (Parslow, Jorm, Christensen, Rodgers, & Jacomb, 2005; Winefield, Black, & Chur-Hansen, 2008). That said, inadequate consideration of potential confounders (Westgarth et al., 2010, 2013) and other design problems may help to explain inconclusive results in human–animal interaction (HAI) studies (Herzog, 2011). Conflicting results from similar studies and failure to replicate findings are especially prevalent in areas of science in which studies are characterized by small and homogenous samples, a wide diversity of research designs, and small effect sizes, all of which apply to this field of research (Ioannidis, 2005, as cited regarding HAI research in Herzog, 2011). Pet owners are often treated as one homogenous population without consideration of differences in their attitudes to pet ownership and pet attachment that are likely to impact potential benefits from their interaction with their pets. Individual differences such as financial resources or health may relate to one's likelihood of becoming a pet owner, and other socioeconomic and demographic factors may be associated with the benefits and costs of pet ownership (Müllersdorf, Granström, Sahlqvist, & Tillgren, 2010; Westgarth et al., 2010).

Nevertheless, pet ownership often leads to increased exercise and reduced anxiety, while providing an external focus of attention that can be beneficial to health (Jennings, 1997; Serpell, 1991). Pet owners fare better in terms of well-being, self-esteem, physical fitness, conscientiousness, extraversion, and attachment security and experience less loneliness and fearfulness than those who don't own pets (McConnell, Brown, Shoda, Stayton, & Martin, 2011). Dogs can be beneficial to individuals with disabilities. In one randomized controlled clinical trial, some individuals with ambulatory disabilities given service dogs had improved in self-esteem and well-being after 6 months compared with a waitlist-control group (Allen & Blascovich, 1996). Also, pets may serve as a protective factor against sickness and disease. Men with AIDS who owned pets reported less depression than men with AIDS who did not; the benefit occurred mainly among men who reported having few confidants (Siegel, 1990). Enhancing felt companionship may buffer against the stressful impacts of disease. Elderly individuals who own pets are less likely to seek physician care, and pet ownership may moderate the relationship between stressful life events and physician contacts (Siegel, 1990). Activities of daily living may deteriorate at a slower rate for elderly individuals who own pets because dog ownership also appears to facilitate mobility (Raina, Waltner-Toews, Bonnett, Woodward, & Abernathy, 1999; Thorpe et al., 2006), and moderate association has been seen between dog ownership and higher levels of walking and physical activity (Yabroff, Troiano, & Berrigan, 2008). Pet owners also are significantly less likely to die within one year of having a heart attack (Friedmann & Thomas, 1995).

SOME TENTATIVE CONCLUSIONS

Although physical and mental health benefits of social relationships and rich social networks for humans are clear, the causes of these associations are not. Perceptions of the environment are influenced by the anticipatory budgeting of resources to promote economy of action. Humans have evolved in social environments, adaptively experiencing energy benefits from social relationships when regulating responses to threatening, stressful, and emotional stimuli. Humans are adapted to function optimally in a social ecology. Social proximity constitutes a baseline situation for neural functioning. Prefrontal resources in particular are taxed when individuals are separated from their social resources. Dogs and other companion animals may themselves live within this human social ecology, benefitting from the human environment. A mutually beneficial relationship is likely, where humans benefit from the presence of companion animals in this ecology. It is likely that animals can fill the role of a social surrogate and confer many benefits of social relationships. Although we suspect that the presence of a companion animal may decrease neural threat responding much like a high-quality relational partner, more HAI research is needed to support such claims. We speculate that the companionship and support pets provide mirrors many of the elements of human relationships that contribute to positive health outcomes.

REFERENCES

Allen, K., & Blascovich, J. (1996). The value of service dogs for people with severe ambulatory disabilities: A randomized controlled trial. JAMA, 275, 1001–1006. http://dx.doi.org/10.1001/jama.1996.03530370039028

American Pet Products Association. (2011). *APPA national pet owners survey 2011–2012*. Retrieved from http://www.americanpetproducts.org/press_industry trends.asp

Associated Press. (2010, April 28). *The AP-Petside.com poll*. Retrieved from http://surveys.ap.org/data/Gfk/AP-GfK%20Petside%20Topline%20for%20final%20041410_1st%20release.pdf

Axelsson, E., Ratnakumar, A., Arendt, M. L., Maqbool, K., Webster, M. T., Perloski, M., . . . Lindblad-Toh, K. (2013). The genomic signature of dog domestication reveals adaptation to a starch-rich diet. *Nature, 495*, 360–364. http://dx.doi.org/10.1038/nature11837

Baumeister, R. F., Bratslavsky, E., Muraven, M., & Tice, D. M. (1998). Ego depletion: Is the active self a limited resource? *Journal of Personality and Social Psychology, 74*, 1252–1265. http://dx.doi.org/10.1037/0022-3514.74.5.1252

Baumeister, R. F., & Vohs, K. D. (2007). Self regulation, ego depletion, and motivation. *Social and Personality Psychology Compass, 1*, 115–128. http://dx.doi.org/10.1111/j.1751-9004.2007.00001.x

Beals, K. P., Peplau, L. A., & Gable, S. L. (2009). Stigma management and well-being: The role of perceived social support, emotional processing, and suppression. *Personality and Social Psychology Bulletin, 35*, 867–879. http://dx.doi.org/10.1177/0146167209334783

Beckes, L., & Coan, J. A. (2011). Social baseline theory: The role of social proximity in emotion and economy of action. *Social and Personality Psychology Compass, 5*, 976–988.

Berscheid, E. (2003). The human's greatest strength: Other humans. In L. G. Aspinwall & U. M. Staudinger (Eds.), *A psychology of human strengths: Fundamental questions and future directions for a positive psychology* (pp. 37–47). San Francisco, CA: Berrett-Kohler.

Bhalla, M., & Proffitt, D. R. (1999). Visual-motor recalibration in geographical slant perception. *Journal of Experimental Psychology: Human Perception and Performance, 25*, 1076–1096. http://dx.doi.org/10.1037/0096-1523.25.4.1076

Bräuer, J., Kaminski, J., Riedel, J., Call, J., & Tomasello, M. (2006). Making inferences about the location of hidden food: Social dog, causal ape. *Journal of Comparative Psychology, 120*, 38–47. http://dx.doi.org/10.1037/0735-7036.120.1.38

Brown, C. L., Oudekerk, B. A., Szwedo, D. E., & Allen, J. P. (2013). Inter-parent aggression as a precursor to disengagement coping in emerging adulthood: The buffering role of friendship competence. *Social Development, 22*, 683–700.

Campos, J. J., Walle, E. A., Dahl, A., & Main, A. (2011). Reconceptualizing emotion regulation. *Emotion Review, 3*, 26–35. http://dx.doi.org/10.1177/1754073910380975

Clutton-Brock, J. (1995). Origins of the dog: Domestication and early history. In J. Serpell (Ed.), *The domestic dog: Its evolution, behavior, and interactions with people* (pp. 7–20). Cambridge, England: Cambridge University Press.

Coan, J. A., Schaefer, H. S., & Davidson, R. J. (2006). Lending a hand: Social regulation of the neural response to threat. *Psychological Science, 17*, 1032–1039.

Cohen, S., & Janicki-Deverts, D. (2009). Can we improve our physical health by altering our social networks? *Perspectives on Psychological Science, 4*, 375–378. http://dx.doi.org/10.1111/j.1745-6924.2009.01141.x

Cooper, J. J., Ashton, C., Bishop, S., West, R., Mills, D. S., & Young, R. J. (2003). Clever hounds: Social cognition in the domestic dog (*Canis familiaris*). *Applied Animal Behaviour Science, 81*, 229–244. http://dx.doi.org/10.1016/S0168-1591(02)00284-8

Coppinger, R., & Coppinger, L. (2001). *Dogs: A startling new understanding of canine origin, behavior and evolution.* New York, NY: Scribner.

Coppola, C. L., Grandin, T., & Enns, R. M. (2006). Human interaction and cortisol: Can human contact reduce stress for shelter dogs? *Physiology & Behavior, 87*, 537–541. http://dx.doi.org/10.1016/j.physbeh.2005.12.001

Derrick, J. L., Gabriel, S., & Hugenberg, K. (2009). Social surrogacy: How favored television programs provide the experience of belonging. *Journal of Experimental Social Psychology, 45*, 352–362. http://dx.doi.org/10.1016/j.jesp.2008.12.003

Dietrich, A., & Sparling, P. B. (2004). Endurance exercise selectively impairs prefrontal-dependent cognition. *Brain and Cognition, 55*, 516–524. http://dx.doi.org/10.1016/j.bandc.2004.03.002

Dunbar, R. I. (1998). The social brain hypothesis. *Brain: A Journal of Neurology, 9*, 10.

Eisenberger, N. I., Master, S. L., Inagaki, T. K., Taylor, S. E., Shirinyan, D., Lieberman, M. D., & Naliboff, B. D. (2011). Attachment figures activate a safety signal-related neural region and reduce pain experience. *Proceedings of the National Academy of Sciences, USA, 108*, 11721–11726. http://dx.doi.org/10.1073/pnas.1108239108

Ellegren, H. (2005). Genomics: The dog has its day. *Nature, 438*, 745–746. http://dx.doi.org/10.1038/438745a

Epley, N., Akalis, S., Waytz, A., & Cacioppo, J. T. (2008). Creating social connection through inferential reproduction: Loneliness and perceived agency in gadgets, gods, and greyhounds. *Psychological Science, 19*, 114–120. http://dx.doi.org/10.1111/j.1467-9280.2008.02056.x

Epley, N., Waytz, A., & Cacioppo, J. T. (2007). On seeing human: A three-factor theory of anthropomorphism. *Psychological Review, 114*, 864–886. http://dx.doi.org/10.1037/0033-295X.114.4.864

Frank, H., & Frank, M. G. (1985). Comparative manipulation-test performance in ten-week-old wolves (*Canis lupus*) and Alaskan malamutes (*Canis familiaris*): A piagetian interpretation. *Journal of Comparative Psychology, 99*, 266–274. http://dx.doi.org/10.1037/0735-7036.99.3.266

Frank, H., Frank, M. G., Hasselbach, L. M., & Littleton, D. M. (1989). Motivation and insight in wolf (*Canis lupus*) and Alaskan malamute (*Canis familiaris*): Visual discrimination learning. *Bulletin of the Psychonomic Society, 27*, 455–458. http://dx.doi.org/10.3758/BF03334654

Friedmann, E., Katcher, A. H., Thomas, S. A., Lynch, J. J., & Messent, P. R. (1983). Social interaction and blood pressure. Influence of animal companions. *Journal of Nervous and Mental Disease, 171*, 461–465. http://dx.doi.org/10.1097/00005053-198308000-00002

Friedmann, E., & Thomas, S. A. (1995). Pet ownership, social support, and one-year survival after acute myocardial infarction in the Cardiac Arrhythmia Suppression Trial (CAST). *The American Journal of Cardiology, 76*, 1213–1217. http://dx.doi.org/10.1016/S0002-9149(99)80343-9

Gailliot, M. T., & Baumeister, R. F. (2007). The physiology of willpower: Linking blood glucose to self-control. *Personality and Social Psychology Review, 11*, 303–327. http://dx.doi.org/10.1177/1088868307303030

Gailliot, M. T., Baumeister, R. F., DeWall, C. N., Maner, J. K., Plant, E. A., Tice, D. M., . . . Schmeichel, B. J. (2007). Self-control relies on glucose as a limited energy source: Willpower is more than a metaphor. *Journal of Personality and Social Psychology, 92*, 325–336. http://dx.doi.org/10.1037/0022-3514.92.2.325

Gallagher, E. N., & Vella-Brodrick, D. A. (2008). Social support and emotional intelligence as predictors of subjective well-being. *Personality and Individual Differences, 44*, 1551–1561. http://dx.doi.org/10.1016/j.paid.2008.01.011

Gibson, J. J. (1986). *The ecological approach to visual perception*. Hillsdale, NJ: Erlbaum.

Gross, J. J., Richards, J. M., & John, O. P. (2006). Emotion regulation in everyday life. In D. K. Snyder, J. A. Simpson, & J. N. Hughes (Eds.), *Emotion regulation in couples and families: Pathways to dysfunction and health* (pp. 13–35). Washington, DC: American Psychological Association. http://dx.doi.org/10.1037/11468-001

Gross, J. J., & Thompson, R. A. (2007). Emotion regulation: Conceptual foundations. In J. J. Gross (Ed.), *Handbook of emotion regulation* (pp. 3–26). New York, NY: Guilford Press.

Haga, S. M., Kraft, P., & Corby, E. K. (2009). Emotion regulation: Antecedents and well-being outcomes of cognitive reappraisal and expressive suppression in cross-cultural samples. *Journal of Happiness Studies, 10*, 271–291. http://dx.doi.org/10.1007/s10902-007-9080-3

Hagger, M. S., & Chatzisarantis, N. L. (2013). The sweet taste of success: The presence of glucose in the oral cavity moderates the depletion of self-control resources. *Personality and Social Psychology Bulletin, 39*, 28–42. http://dx.doi.org/10.1177/0146167212459912

Hare, B., Brown, M., Williamson, C., & Tomasello, M. (2002). The domestication of social cognition in dogs. *Science, 298*, 1634–1636. http://dx.doi.org/10.1126/science.1072702

Hare, B., & Tomasello, M. (2005). Human-like social skills in dogs? *Trends in Cognitive Sciences, 9*, 439–444. http://dx.doi.org/10.1016/j.tics.2005.07.003

Hennessy, M. B., Williams, M. T., Miller, D. D., Douglas, C. W., & Voith, V. L. (1998). Influence of male and female petters on plasma cortisol and behaviour: Can human interaction reduce the stress of dogs in a public animal shelter? *Applied Animal Behaviour Science, 61*, 63–77. http://dx.doi.org/10.1016/S0168-1591(98)00179-8

Herzog, H. (2011). The impact of pets on human health and psychological well-being: Fact, fiction, or hypothesis? *Current Directions in Psychological Science, 20*, 236–239. http://dx.doi.org/10.1177/0963721411415220

Holt-Lunstad, J., Smith, T. B., & Layton, J. B. (2010). Social relationships and mortality risk: A meta-analytic review. *PLoS Medicine, 7*, e1000316. http://dx.doi. org/10.1371/journal.pmed.1000316

House, J. S., Landis, K. R., & Umberson, D. (1988). Social relationships and health. *Science, 241*, 540–545. http://dx.doi.org/10.1126/science.3399889

Humane Society of the United States. (2011). *U.S. pet ownership statistics.* Retrieved from http://www.humanesociety.org/issues/pet_overpopulation/facts/pet_ ownership_statistics.html

Ioannidis, J. P. A. (2005). Why most published research findings are false. *PLoS Medicine, 2*, e124. http://dx.doi.org/10.1371/journal.pmed.0020124

Jennings, L. B. (1997). Potential benefits of pet ownership in health promotion. *Journal of Holistic Nursing, 15*, 358–372. http://dx.doi.org/10.1177/ 089801019701500404

Job, V., Dweck, C. S., & Walton, G. M. (2010). Ego depletion—is it all in your head? Implicit theories about willpower affect self-regulation. *Psychological Science, 21*, 1686–1693. http://dx.doi.org/10.1177/0956797610384745

Kurzban, R. (2010). Does the brain consume additional glucose during self-control tasks? *Evolutionary Psychology, 8*, 244–259. http://dx.doi.org/10.1177/ 147470491000800208

Lane, D. R., McNicholas, J., & Collis, G. M. (1998). Dogs for the disabled: Benefits to recipients and welfare of the dog. *Applied Animal Behaviour Science, 59*, 49–60. http://dx.doi.org/10.1016/S0168-1591(98)00120-8

Lopes, P. N., Salovey, P., Côté, S., Beers, M., & Petty, R. E. (2005). Emotion regulation abilities and the quality of social interaction. *Emotion, 5*, 113–118. http:// dx.doi.org/10.1037/1528-3542.5.1.113

Lynch, J. J., & McCarthy, J. F. (1969). Social responding in dogs: Heart rate changes to a person. *Psychophysiology, 5*, 389–393. http://dx.doi.org/10.1111/j.1469-8986.1969. tb02838.x

McConnell, A. R., Brown, C. M., Shoda, T. M., Stayton, L. E., & Martin, C. E. (2011). Friends with benefits: On the positive consequences of pet ownership. *Journal of Personality and Social Psychology, 101*, 1239–1252. Advance online publication. http://dx.doi.org/10.1037/a0024506

McNicholas, J., & Collis, G. M. (2000). Dogs as catalysts for social interactions: Robustness of the effect. *British Journal of Psychology, 91*, 61–70. http://dx.doi.org/ 10.1348/000712600161673

McNicholas, J., Gilbey, A., Rennie, A., Ahmedzai, S., Dono, J. A., & Ormerod, E. (2005). Pet ownership and human health: A brief review of evidence and issues. *British Medical Journal, 331*, 1252–1254. http://dx.doi.org/10.1136/ bmj.331.7527.1252

Mersmann, D., Tomasello, M., Call, J., Kaminski, J., & Taborsky, M. (2011). Simple mechanisms can explain social learning in domestic dogs (*Canis familiaris*). *Ethology, 117*, 675–690. http://dx.doi.org/10.1111/j.1439-0310.2011.01919.x

Miklósi, Á., Topál, J., & Csányi, V. (2004). Comparative social cognition: What can dogs teach us? *Animal Behaviour, 67,* 995–1004. http://dx.doi.org/10.1016/j.anbehav.2003.10.008

Mobbs, D., Petrovic, P., Marchant, J. L., Hassabis, D., Weiskopf, N., Seymour, B., . . . Frith, C. D. (2007). When fear is near: Threat imminence elicits prefrontal-periaqueductal gray shifts in humans. *Science, 317,* 1079–1083. http://dx.doi.org/10.1126/science.1144298

Müllersdorf, M., Granström, F., Sahlqvist, L., & Tillgren, P. (2010). Aspects of health, physical/leisure activities, work and socio-demographics associated with pet ownership in Sweden. *Scandinavian Journal of Public Health, 38,* 53–63. http://dx.doi.org/10.1177/1403494809344358

Noonan, G. J., Rand, J. S., Blackshaw, J. K., & Priest, J. (1996). Tail docking in dogs: A sample of attitudes of veterinarians and dog breeders in Queensland. *Australian Veterinary Journal, 73*(3), 86–88. http://dx.doi.org/10.1111/j.1751-0813.1996.tb09982.x

Parslow, R. A., Jorm, A. F., Christensen, H., Rodgers, B., & Jacomb, P. (2005). Pet ownership and health in older adults: Findings from a survey of 2,551 community-based Australians aged 60–64. *Gerontology, 51,* 40–47. http://dx.doi.org/10.1159/000081433

Phelps, E. A., & LeDoux, J. E. (2005). Contributions of the amygdala to emotion processing: From animal models to human behavior. *Neuron, 48,* 175–187.

Pongrácz, P., Miklósi, Á., Kubinyi, E., Gurobi, K., Topál, J., & Csányi, V. (2001). Social learning in dogs: The effect of a human demonstrator on the performance of dogs in a detour task. *Animal Behaviour, 62,* 1109–1117. http://dx.doi.org/10.1006/anbe.2001.1866

Proffitt, D. R. (2006). Embodied perception and the economy of action. *Perspectives on Psychological Science, 1,* 110–122. http://dx.doi.org/10.1111/j.1745-6916.2006.00008.x

Proffitt, D. R., Bhalla, M., Gossweiler, R., & Midgett, J. (1995). Perceiving geographical slant. *Psychonomic Bulletin & Review, 2,* 409–428. http://dx.doi.org/10.3758/BF03210980

Raina, P., Waltner-Toews, D., Bonnett, B., Woodward, C., & Abernathy, T. (1999). Influence of companion animals on the physical and psychological health of older people: An analysis of a one-year longitudinal study. *Journal of the American Geriatrics Society, 47,* 323–329. http://dx.doi.org/10.1111/j.1532-5415.1999.tb02996.x

Reid, P. J. (2009). Adapting to the human world: Dogs' responsiveness to our social cues. *Behavioural Processes, 80,* 325–333. http://dx.doi.org/10.1016/j.beproc.2008.11.002

Richerson, P. J., Boyd, R., & Henrich, J. (2010). Gene-culture coevolution in the age of genomics. *Proceedings of the National Academy of Sciences, USA, 107*(Suppl. 2), 8985–8992. http://dx.doi.org/10.1073/pnas.0914631107

Schnall, S., Harber, K. D., Stefanucci, J. K., & Proffitt, D. R. (2008). Social support and the perception of geographical slant. *Journal of Experimental Social Psychology, 44*, 1246–1255. http://dx.doi.org/10.1016/j.jesp.2008.04.011

Schnall, S., Zadra, J. R., & Proffitt, D. R. (2010). Direct evidence for the economy of action: Glucose and the perception of geographical slant. *Perception, 39*, 464–482. http://dx.doi.org/10.1068/p6445

Serpell, J. (1991). Beneficial effects of pet ownership on some aspects of human health and behaviour. *Journal of the Royal Society of Medicine, 84*, 717–720.

Siegel, J. M. (1990). Stressful life events and use of physician services among the elderly: The moderating role of pet ownership. *Journal of Personality and Social Psychology, 58*, 1081–1086. http://dx.doi.org/10.1037/0022-3514.58.6.1081

Smyth, J. M., & Arigo, D. (2009). Recent evidence supports emotion-regulation interventions for improving health in at-risk and clinical populations. *Current Opinion in Psychiatry, 22*, 205–210. http://dx.doi.org/10.1097/YCO.0b013e3283252d6d

Stafford, K. (2007). *The welfare of dogs*. Dordrecht, The Netherlands: Springer.

Thorpe, R. J., Jr., Simonsick, E. M., Brach, J. S., Ayonayon, H., Satterfield, S., Harris, T. B., . . . Kritchevsky, S. B., for the Health, Aging and Body Composition Study. (2006). Dog ownership, walking behavior, and maintained mobility in late life. *Journal of the American Geriatrics Society, 54*, 1419–1424. http://dx.doi.org/10.1111/j.1532-5415.2006.00856.x

Tomasello, M., & Carpenter, M. (2007). Shared intentionality. *Developmental Science, 10*, 121–125. http://dx.doi.org/10.1111/j.1467-7687.2007.00573.x

Topál, J., Byrne, R. W., Miklósi, Á., & Csányi, V. (2006). Reproducing human actions and action sequences: "Do as I Do!" in a dog. *Animal Cognition, 9*, 355–367. http://dx.doi.org/10.1007/s10071-006-0051-6

Topál, J., Miklósi, Á., & Csányi, V. (1997). Dog–human relationship affects problem solving behaviour in dogs. *Anthrozoös, 10*, 214–224. http://dx.doi.org/10.2752/089279397787000987

Topál, J., Miklósi, Á., Csányi, V., & Dóka, A. (1998). Attachment behavior in dogs (*Canis familiaris*): A new application of Ainsworth's (1969) Strange Situation Test. *Journal of Comparative Psychology, 112*, 219–229. http://dx.doi.org/10.1037/0735-7036.112.3.219

Uchino, B. N., Cacioppo, J. T., & Kiecolt-Glaser, J. K. (1996). The relationship between social support and physiological processes: A review with emphasis on underlying mechanisms and implications for health. *Psychological Bulletin, 119*, 488–531. http://dx.doi.org/10.1037/0033-2909.119.3.488

Udell, M. A. R., Dorey, N. R., & Wynne, C. D. L. (2008). Wolves outperform dogs in following human social cues. *Animal Behaviour, 76*, 1767–1773. http://dx.doi.org/10.1016/j.anbehav.2008.07.028

Udell, M. A., Dorey, N. R., & Wynne, C. D. (2010). What did domestication do to dogs? A new account of dogs' sensitivity to human actions. *Biological Reviews*

of the *Cambridge Philosophical Society, 85,* 327–345. http://dx.doi.org/10.1111/
j.1469-185X.2009.00104.x

Wager, T. D., Davidson, M. L., Hughes, B. L., Lindquist, M. A., & Ochsner, K. N.
(2008). Prefrontal-subcortical pathways mediating successful emotion regulation. *Neuron, 59,* 1037–1050. http://dx.doi.org/10.1016/j.neuron.2008.09.006

Westgarth, C., Boddy, L. M., Stratton, G., German, A. J., Gaskell, R. M., Coyne,
K. P., . . . Dawson, S. (2013). Pet ownership, dog types and attachment to pets
in 9–10 year old children in Liverpool, UK. *BMC Veterinary Research, 9*(1), 102.
http://dx.doi.org/10.1186/1746-6148-9-102

Westgarth, C., Heron, J., Ness, A. R., Bundred, P., Gaskell, R. M., Coyne, K. P., . . .
Dawson, S. (2010). Family pet ownership during childhood: Findings from a UK
birth cohort and implications for public health research. *International Journal of
Environmental Research and Public Health, 7,* 3704–3729. http://dx.doi.org/10.3390/
ijerph7103704

Williams, L. E., & Bargh, J. A. (2008). Experiencing physical warmth promotes
interpersonal warmth. *Science, 322*(5901), 606–607. http://dx.doi.org/10.1126/
science.1162548

Winefield, H. R., Black, A., & Chur-Hansen, A. (2008). Health effects of ownership of and attachment to companion animals in an older population.
International Journal of Behavioral Medicine, 15, 303–310. http://dx.doi.org/
10.1080/10705500802365532

Yabroff, K. R., Troiano, R. P., & Berrigan, D. (2008). Walking the dog: Is pet ownership associated with physical activity in California? *Journal of Physical Activity
& Health, 5,* 216–228.

Zadra, J. R., & Clore, G. L. (2011). Emotion and perception: The role of affective information. *Wiley Interdisciplinary Reviews: Cognitive Science, 2,* 676–685.
http://dx.doi.org/10.1002/wcs.147

7

UNDERSTANDING EMPATHY AND PSYCHOPATHY THROUGH COGNITIVE AND SOCIAL NEUROSCIENCE

LEAH M. LOZIER, KRISTIN M. BRETHEL-HAURWITZ,
AND ABIGAIL A. MARSH

Social relationships between humans and domesticated animals can often be mutually beneficial and affectionate. Humans and domestic animals provide each other with various instrumental benefits, including food provision, protection, and play. But humans' relationships with domesticated animals are not simply instrumental. Interactions of humans with their pets, and even with unfamiliar animals, are often marked by warmth, affection, and empathy. The capacity for humans to exhibit empathy toward nonhuman animals represents something of a puzzle. *Empathy*, a term that refers to a constellation of phenomena, including the capacity to know or understand the contents of another individual's mind and the capacity to care about another individual's well-being (de Waal, 2008; Decety, 2015), is generally believed to be more commonly experienced toward similar individuals than toward dissimilar individuals (Krebs, 1975). That humans can experience empathy for nonhuman animals that are extremely unlike

http://dx.doi.org/10.1037/14856-009
The Social Neuroscience of Human–Animal Interaction, L. S. Freund, S. McCune, L. Esposito, N. R. Gee, and P. McCardle (Editors)

147

themselves indicates that important dynamics other than similarity must be in play. Considering the possibility of empathy across species may aid in understanding both the dynamics of empathy and of human–animal interactions (HAIs). This chapter considers the various forms of empathy, explores the neural bases of those forms most relevant to compassionate social behavior within and across species, and discusses the consequences that may result from empathy impairments.

The study of empathy has become increasingly popular among psychology and neuroscience researchers. However, research on empathy is often hampered by the use of the term *empathy* to describe a variety of overlapping but distinct phenomena, including cognitive perspective taking, sympathetic concern, and emotional contagion. An exploration of empathy as it relates to important behavioral phenomena, including aggression, altruism, and HAIs, requires that care be taken to distinguish among these phenomena.

Cognitive perspective taking (sometimes called *theory of mind*) is the most cognitively complex form of empathy (de Waal, 2008). Perspective taking, the process of acquiring insight into another person's intentions, desires, and beliefs via attempts to infer or adopt his or her point of view (Baron-Cohen, 1997), is often described colloquially as putting oneself into another's shoes. It is typically achieved by considering prior knowledge of the person and his or her current context. For example, if I know that a friend is hoping to adopt a companion dog and I am in the room when he takes a call from the agency handling rescue adoptions, I can infer that he will not want me to make any noise or disturb him while he takes the call. The sophisticated processes that compose perspective taking are essential for many higher level social functions, such as understanding deception (Hala, Chandler, & Fritz, 1991). Impaired perspective taking characterizes some neurodevelopmental disorders that result in profound social functioning deficits, most notably schizophrenia and autism (Crespi & Badcock, 2008).

Sympathetic concern is the form of empathy typically associated with feelings of sorrow, compassion, or pity for the suffering of another being (de Waal, 2008; Nichols, 2001). For example, if a friend or your dog were injured and showing signs of distress, you might experience sympathetic concern as a result. Sympathetic concern entails two distinct phenomena: being aware of another person's distress and desiring to help ameliorate the distress. It is particularly this latter element—the desire to help—that is difficult to capture using currently available methods. In general, self-report is the predominant means of assessing sympathetic concern. Self-report is an essential means of acquiring psychological information, but this method has a variety of limitations and shortcomings that often must be mitigated by, for example, enabling anonymous responding, or by supplementing self-report with complementary information-gathering strategies such psychophysiological or neurophysiological measurements (Brener, Billy, & Grady, 2003).

Alternatively, some of the difficulties inherent in measuring sympathetic concern can be ameliorated by focusing on the first component of this phenomenon: the simple awareness or detection of another person's distress. This most basic form of empathy is sometimes referred to as *emotional contagion*, a term that suggests emotions may spread among individuals through very low-level processes outside of conscious awareness (de Waal, 2008). Because it is relatively simple, this is a form of empathy that is much more amenable to multiple types of measurement. The detection of another person's distress can be measured by asking study participants to label emotion expressed in another's face, voice, or body or by using psychophysiological or neurophysiological measurements to identify changes in the participant's nervous system that indicate detection of another's distress. So, for example, we could infer that you experienced emotional contagion in response to a human or a dog who was injured if you were able to correctly label that individual's nonverbal cues as signs of distress or if viewing these cues caused you to exhibit changes in autonomic arousal such as alterations in heart rate or respiration.

RESPONDING TO DISTRESS

Responses to nonverbal distress cues are particularly illuminating for the study of empathy, as they appear to promote both emotional contagion and sympathetic concern and therefore enhance affiliative, prosocial responses. For example, fearful facial expressions, which convey acute and urgent distress, elicit self-reported sympathetic concern from perceivers (Marsh & Ambady, 2007) as well as promoting affiliative behaviors such as behavioral approach. Approach can be measured using a lever and comparing the response times required to push versus pull the lever in response to the presentation of a stimulus (Hammer & Marsh, 2015; Marsh, Ambady, & Kleck, 2005). That fearful expressions consistently elicit more rapid pulling (approach) than pushing (avoidance) is consistent with their being perceived as appetitive stimuli, despite fear typically being considered a "negative" emotion (Izard, 2007).

To understand why fearful facial expressions and other distress cues elicit sympathetic concern and behavioral approach, it may help to consider how fear-related behaviors are used in other social species, such as wolves and domestic dogs. Species-specific fear behaviors displayed by wolves and dogs when they anticipate attack from another wolf or dog include crouching, tucking the tail, pinning back the ears, licking the antagonist's jowls, and rolling on the back (Schenkel, 1967). The primary purpose of these cues is to signify submission—that the expresser does not represent a threat or competition to the antagonist (Schenkel, 1967; Smith & Price, 1973). These stereotyped behaviors are thought to serve this function by making the fearful, submissive wolf or dog appear smaller and more helpless by mimicking appearance cues of pups (Schenkel, 1967). In fact, across multiple social species, it is thought that fear and submission cues mimic infantile cues because adults typically inhibit aggression toward juveniles (Lorenz, 1966).

Considering fear across species permits the generation of a hypothesis for the social function served by human expressions of fear: to appear submissive or infantile, thereby inhibiting aggression or eliciting help from perceivers. This hypothesis is consistent with findings that fearful expressions appear affiliative, as a babyish facial appearance is consistently perceived as highly affiliative (Zebrowitz & Montepare, 1992), and with the fact that fearful expressions elicit approach, as this is a behavioral response associated with protective behaviors like nurturing the young (Burgdorf & Panksepp, 2006). It is also consistent with findings that fearful expressions are perceived as looking babyish by observers, even when the expressions are manipulated to prevent observers from identifying the emotion being expressed (Marsh, Adams, & Kleck, 2005). This finding suggests that the appearance of the features in a fearful face—the widened eyes, raised brows, flattened brow ridge—makes the expresser appear more babyish physically. Fearful expressions may serve appeasing functions by eliciting the same responses that people have toward infants, just as fear and submission cues in wolves and other social species do (Hammer & Marsh, 2015).

Together, these lines of evidence may help clarify why normal responses to fear are associated with empathy, as has been demonstrated by several paradigms. People report experiencing high levels of sympathetic concern in response to fearful faces, even if the fearful faces are presented subliminally and observers have no conscious awareness of having seen them (Marsh & Ambady, 2007). Moreover, the extent to which people feel sympathetic concern in response to others' fearful expressions is associated with how sympathetic those people are in general. People who recognize fearful expressions better are more likely to respond altruistically in response to others' distress (Marsh, Kozak, & Ambady, 2007). It has been theorized that perhaps the most important predictor of the capacity for sympathetic concern in response to others' distress is the simple capacity to correctly recognize others' distress (Nichols, 2001). This capacity may extend to concern for other species' distress as well; in one recent study, approach behavior was consistently observed in mother deer exposed to the distress vocalizations of several species of mammalian infants, including human infants (Lingle & Riede, 2014).

THE NEURAL BASIS OF EMPATHY

An understanding of the neural processes that underlie any psychological phenomenon can be acquired in a variety of ways. Neural functioning in healthy and atypical brains can be measured and compared, an approach that has been enabled by the development of novel neuroimaging techniques in recent decades. Human lesion cases can also be used to investigate neurocognitive deficits that arise following injury to specific structures. For example, it has been shown that deficits directly relevant to impaired empathic functioning, including impaired recognition of and responses to distress cues, arise in individuals with damage to the amygdala.

The amygdala is a subcortical structure located within the temporal lobe that comprises several subnuclei. The lateral and basolateral nuclei of the amygdala receive sensory information via the thalamus and cortex, and relay this information to the central nucleus of the amygdala. The central nucleus projects to other subcortical regions, including the lateral hypothalamus and regions of the brain stem, which directly mediate physiological and behavioral aspects of fear responding, such as changes in heart rate, skin conductance, and freezing or flight behavior (Davis & Whalen, 2001). Reciprocal connections between the nuclei of the amygdala and the orbitofrontal cortex, hippocampus, and striatum mediate other processes central to fear responding, including fear learning and memory and instrumental approach or avoidance behavior (Davis & Whalen, 2001; LeDoux, 2003; Maren, 2001). That intact amygdala function underlies fearful responding to incipient threats has been established from numerous animal studies—in, for example, rats, rabbits, cats, and dogs—documenting reduced fear responses following amygdala damage and heightened fear responses following amygdala stimulation (Davis, 1992; Goddard, 1964).

More recent research on humans with focal amygdala lesions has established that the amygdala is important for the recognition of others' fear responses as well. One of the best-studied lesion patients is SM (Adolphs, Tranel, Damasio, & Damasio, 1994), a woman with Urbach-Wiethe disease that caused selective and almost complete calcification of her amygdalae. Repeated testing found SM to exhibit marked deficits in recognizing fearful facial expressions, although her recognition of other emotional expressions like happiness, sadness, and disgust was largely intact. By contrast, she showed no impairment in other aspects of face processing, such as recognition of facial identity (Adolphs et al., 1994).

Although very few individuals show both complete and selective amygdala damage akin to SM's, research on other patients with lesions that include the amygdala, either unilaterally or bilaterally, generally has confirmed that individuals with amygdala lesions are impaired in recognizing fear expressed via the face, body, or voice (Adolphs et al., 1999; Scott et al., 1997; Sprengelmeyer et al., 1999).

PSYCHOPATHY AS A CLINICAL MODEL OF IMPAIRED EMPATHY

Amygdala dysfunction may be still more pernicious when it is developmental. Whereas most adults with amygdala lesions acquired these lesions in adolescence or adulthood, some developmental disorders also impair the functioning of the amygdala. One notable example of such a disorder is psychopathy. This disorder is of particular importance to understanding empathy because of the suite of impairments typically observed among individuals with psychopathic personality traits, which include impaired amygdala functioning; impaired ability to recognize and respond to nonverbal distress cues, such as fearful facial expressions; and persistent and severe antisocial behavior, including aggression toward both humans and nonhuman animals.

Psychopathy is characterized by emotional deficits such as shallow affect; difficulty forming close bonds with others; a lack of remorse or empathy; a parasitic orientation toward others; irresponsible and disinhibited behavior; and goal-directed aggression, such as threats, intimidation, and physical violence (Hare, 1991; Skeem, Polaschek, Patrick, & Lilienfeld, 2011). Accumulating social neuroscience literature links psychopathy to characteristic abnormalities in neural structure and functioning in both clinical and community samples (Blair, Finger, & Marsh, 2009).

These neural abnormalities may also represent a risk factor for cruelty toward nonhuman animals, which may be a specific early indicator of psychopathy (Dadds, Whiting, & Hawes, 2006). Animal abuse has long been associated with patterns of generally disruptive and violent behavior, such as violence toward family members. In one recent notorious case, a 19-year-old Washington State man responded to his mother's threat to cut off his allowance by butchering his family's two dogs, then locking his mother and grandmother inside their home, describing what he had done to the dogs, and threatening to kill his mother and grandmother as well unless they agreed to keep paying his allowance (Cuniff, 2010). Individuals who engage in this type of behavior are at risk for more serious behavior disruptions and worse long-term outcomes than are youths with conduct problems who do not abuse nonhuman animals (Luk, Staiger, Wong, & Mathai, 1999). This suggests that cruelty toward nonhuman animals may result from risk factors beyond the family conflicts and parenting problems that place children at risk for generally disruptive behavior. One recent study suggests that psychopathy may be a key factor in the emergence of animal cruelty (Dadds et al., 2006). In this study, affective traits associated with psychopathy, such as reduced empathy, predicted animal abuse better than overall levels of aggressive behavior or levels of family conflict. This suggests that animal cruelty may represent an early indicator that may later manifest as aggression and violence toward humans. Ultimately, then, a more complete understanding of psychopathy may be important for understanding animal abuse as well as for understanding empathy more generally.

Psychopathy is not associated with impairments in all forms of empathy. Individuals who are psychopathic frequently show intact perspective taking (Blair, 2008b), suggesting that perspective taking (or lack thereof) may not be tightly linked to callousness or the types of aggression most closely linked to psychopathy. De Waal (2008) suggested as much, proposing that, "Without emotional engagement . . . perspective taking would be a cold phenomenon that could just as easily lead to torture as to helping" (p. 287). To take the earlier example, simply knowing that a friend will not want you to disturb him may or may not motivate you to act as he wishes. Inferring the nature of others' internal states is not identical to caring about whether those states are good or bad. Indeed, psychopathic individuals may rely on intact perspective-taking abilities to con, manipulate, and otherwise take advantage of others.

By contrast, responses to others' distress, which as mentioned previously are indicative of the low-level form of empathy known as emotional contagion, are unquestionably deficient in psychopathy. In response to depictions of others' distress, individuals with psychopathic traits are less likely to show psychophysiological or neurophysiological changes (Blair, 1999) associated with emotion perception than typical individuals (Blair, 1999; Jones, Laurens, Herba, Barker, & Viding, 2009; Marsh et al., 2008; White et al., 2012). They are also impaired in recognizing distress expressed via the face, body, or voice (Marsh & Blair, 2008; Muñoz, 2009). Moreover, accumulating data suggest that empathic deficits in psychopathy are particularly evident in response to fearful facial expressions. A meta-analysis by Marsh and Blair (2008) found largely intact facial expression recognition abilities for emotions like anger, disgust, happiness, and surprise in participants who were psychopathic or otherwise antisocial, but found that these individuals were roughly 20% worse than control participants at recognizing facial expressions of fear. In other words, the ability to simply recognize others' distress, perhaps the most fundamental form of empathy, is compromised in psychopathy.

EMPATHY AND RESPONSES TO FEAR

Returning to a consideration of the role of distress cues in social interactions may help to illuminate the mechanisms by which psychopathy is associated with aggression toward humans and nonhuman animals. Distress cues appear to function across many species, humans included, to promote affiliative, prosocial behavior, and sympathetic concern. That individuals with psychopathy cannot reliably recognize these cues and show aberrant physiological responses to them suggests dysfunctional processing of distress cues in psychopathy, which may impair empathic responses to others' distress (Blair, 1999; 2008b). This may make psychopathic individuals more likely to engage in aggressive interpersonal behaviors that cause others distress (Blair, 2005; Marsh & Cardinale, 2012). A consideration of the neurocognitive impairments associated with psychopathy may help to clarify why this is the case.

NEUROCOGNITIVE IMPAIRMENTS IN PSYCHOPATHY

Behavioral and cognitive similarities between individuals with psychopathy and those with amygdala lesions led to early hypotheses that amygdala dysfunction may underlie core features of psychopathy (Blair, 2003; Blair, Colledge, Murray, & Mitchell, 2001). These hypotheses are supported by recent experiments using structural and functional neuroimaging that suggest atypical amygdala morphology and functional hypoactivation in individuals with psychopathic traits.

A large and well-controlled volumetric study in 296 incarcerated men found that psychopathy was associated with reduced bilateral gray matter volume in the amygdala (Ermer, Cope, Nyalakanti, Calhoun, & Kiehl, 2012). Yang, Raine, Narr, Colletti, and Toga (2009) identified reduced amygdala volume in both the basolateral and central nuclei of individuals with psychopathy. Other studies have found reduced structural integrity in key white matter connections between the amygdala and frontal areas implicated in executive and emotional control (Craig et al., 2009; Motzkin, Newman, Kiehl, & Koenigs, 2011). Although at least one study found evidence for mixed patterns of increased and reduced size within amygdala subnuclei (Boccardi et al., 2011), on the whole, the evidence is largely consistent regarding structural abnormalities in the amygdala in psychopathy.

Given indications that psychopathy impairs recognition of distress cues like fearful expressions and that amygdala damage impairs fear recognition, it follows that individuals with psychopathic traits may exhibit amygdala dysfunction when viewing fearful expressions. The results of several recent neuroimaging studies support this prediction. Both adolescents and adults with psychopathic traits appear to show amygdala hypoactivation in response to fearful expressions but not other expressions (Jones et al., 2009; Lozier, Cardinale, VanMeter, & Marsh, 2014; Marsh et al., 2008; White et al., 2012). In addition, psychopathy is associated with reduced functional connectivity between the amygdala and the prefrontal cortex during expression processing tasks (Marsh et al., 2008), indicating that impaired processing of fearful expressions may have broad implications for learning or behavioral responses associated with these expressions.

It is theorized that dysfunction in the amygdala and the structures with which it is reciprocally connected (particularly orbitofrontal cortex, anterior cingulate cortex, and striatum) can explain not only low-level empathic impairments in psychopathy, such as recognizing and responding to others' distress, but also impairments in higher level socioemotional processes such as cooperation, judgments about aggression, and moral reasoning (Blair, 2008a; Glenn, Raine, & Schug, 2009; Harenski, Harenski, Shane, & Kiehl, 2010; Rilling et al., 2007). For example, Rilling et al. (2007) found that psychopathy was associated with aberrant patterns of activation in amygdala and orbitofrontal cortex during a Prisoner's Dilemma task, in which study participants gain or lose points depending on whether they choose to cooperate with other players. Individuals with psychopathic traits were less likely to cooperate and more likely to defect (choose noncooperation) during the task, and they showed reduced amygdala activation when their partners defected (Rilling et al., 2007). Other studies have found that psychopathy is linked to reduced amygdala and medial prefrontal cortex activation and coactivation during moral reasoning tasks, in which participants typically must make valenced judgments about the severity of immoral behaviors (Harenski et al., 2010; Marsh et al., 2011). Differences between neural responses in psychopathic and control groups may be most likely to emerge when moral judgments are emotional in nature (Glenn et al., 2009).

NEUROCOGNITIVE MODELS OF EMPATHIC DYSFUNCTION IN PSYCHOPATHY

Accumulating research findings exploring the behavioral, physiological, and neural correlates of psychopathy have led to relatively consistent outcomes, and it is widely accepted that psychopathy is associated with dysfunction in amygdala and associated structures such as orbitofrontal cortex, anterior cingulate cortex, and striatum (Blair, 2007; Harenski et al., 2010; Raine, 2008). However, a variety of interpretations of the mechanisms by which dysfunction in these structures leads to the patterns of cognition, emotional responding, and behavioral characteristics of psychopathy have been proposed.

For example, the low-fear hypothesis (Lykken, 1995) posits that psychopathy derives from heritable traits that reduce harm avoidance and increase the penchant for risk taking. However, this model does not directly address the link between fearlessness and low-level empathy and sympathetic concern, the state that motivates people to inhibit aggression or provide help to a distressed individual. Alternatively, the integrated emotion systems model (Blair, 2005) identifies the amygdala as a primary source of dysfunction in psychopathy, focusing on its essential role in aversive conditioning. This model does not focus on empathy per se, although it provides a clear mechanism by which psychopathy would lead to increases in antisocial behavior (and perhaps reductions in prosocial behavior). However, its underlying assumption is that distress cues like fearful facial expressions are unconditioned aversive stimuli, which is inconsistent with evidence that fearful expressions elicit behavioral approach (Marsh, Ambady, & Kleck, 2005). Finally, the response modulation model (Newman, 1998) views distress cues as unconditioned aversive stimuli and posits that psychopathy impedes one's ability to allocate attentional resources to peripheral emotional information if considered irrelevant to current goal-directed behavior. This model suggests that response modulation deficits result in failures to link the affective consequences of antisocial behaviors to the actions themselves (Newman, 1998). However, this model does not adequately reflect dominant models of attention (Blair, Mitchell, & Blair, 2005; Desimone & Duncan, 1995), and recent findings raise doubts that attentional processes are central to emotion processing deficits in psychopathy (Anderson & Stanford, 2012; Sylvers, Brennan, & Lilienfeld, 2011).

As yet, then, no single model of psychopathy captures the totality of the available evidence regarding the neurocognitive and neurobiological features of this disorder. As researchers' understanding of psychopathy and its neurobiological correlates evolves, the models that have been developed to explain features of the disorder will evolve as well. This may enable the improved identification of targets of intervention and treatment for individuals with psychopathy.

CONCLUSION

Individuals with psychopathic traits are dramatically overrepresented among violent criminal offenders (Blair et al., 2005). This fact adds urgency to the ongoing search for explanations of why some individuals engage in cruel behavior toward both humans and nonhuman animals. Low-level deficits in recognizing and responding to the distress of others are central to psychopathy, and this kind of deficit may disrupt the most fundamental form of empathy (de Waal, 2008). Perhaps the similarities between human distress cues and their counterparts among other social species like dogs provide some clue as to why animal cruelty is an apt predictor of aggression toward humans in individuals with psychopathic traits. In social species, cues that signal distress, particularly fear, normally serve to inhibit aggression and elicit nurturing responses among those who see them. But for individuals with psychopathic traits, in whom characteristic patterns of brain dysfunction may have disrupted the ability to understand and respond to these cues, the ability to regulate behavior in response to the fear and distress of others, whether human or canine, may be disrupted as well. Using cognitive neuroscience methods like functional neuroimaging to explore the neurobiological roots of empathy and psychopathy in humans and corresponding behavioral responses in other social species, particularly dogs, may enable us to develop a clearer understanding of these disruptions.

REFERENCES

Adolphs, R., Tranel, D., Damasio, H., & Damasio, A. (1994). Impaired recognition of emotion in facial expressions following bilateral damage to the human amygdala. *Nature, 372,* 669–672. http://dx.doi.org/10.1038/372669a0

Adolphs, R., Tranel, D., Hamann, S., Young, A. W., Calder, A. J., Phelps, E. A., . . . Damasio, A. R. (1999). Recognition of facial emotion in nine individuals with bilateral amygdala damage. *Neuropsychologia, 37,* 1111–1117. http://dx.doi.org/10.1016/S0028-3932(99)00039-1

Anderson, N. E., & Stanford, M. S. (2012). Demonstrating emotional processing differences in psychopathy using affective ERP modulation. *Psychophysiology, 49,* 792–806. Advance online publication. http://dx.doi.org/10.1111/j.1469-8986.2012.01369.x

Baron-Cohen, S. (1997). Mindreading: Nature's choice. In *Mindblindness: An essay on autism and theory of mind* (pp. 21–30). Cambridge, MA: MIT Press.

Blair, R. J. R. (1999). Responsiveness to distress cues in the child with psychopathic tendencies. *Personality and Individual Differences, 27,* 135–145. http://dx.doi.org/10.1016/S0191-8869(98)00231-1

Blair, R. J. (2003). Neurobiological basis of psychopathy. *The British Journal of Psychiatry, 182,* 5–7. http://dx.doi.org/10.1192/bjp.182.1.5

Blair, R. J. (2005). Applying a cognitive neuroscience perspective to the disorder of psychopathy. *Development and Psychopathology, 17,* 865–891. http://dx.doi.org/10.1017/S0954579405050418

Blair, R. J. (2007). The amygdala and ventromedial prefrontal cortex in morality and psychopathy. *Trends in Cognitive Sciences, 11,* 387–392. http://dx.doi.org/10.1016/j.tics.2007.07.003

Blair, R. J. (2008a). The cognitive neuroscience of psychopathy and implications for judgments of responsibility. *Neuroethics, 1,* 149–157. http://dx.doi.org/10.1007/s12152-008-9016-6

Blair, R. J. (2008b). Fine cuts of empathy and the amygdala: Dissociable deficits in psychopathy and autism [Special issue]. *Quarterly Journal of Experimental Psychology, 61,* 157–170. http://dx.doi.org/10.1080/17470210701508855

Blair, R. J., Colledge, E., Murray, L., & Mitchell, D. G. (2001). A selective impairment in the processing of sad and fearful expressions in children with psychopathic tendencies. *Journal of Abnormal Child Psychology, 29,* 491–498. http://dx.doi.org/10.1023/A:1012225108281

Blair, R. J. R., Finger, E., & Marsh, A. A. (2009). The development and neural bases of psychopathy. In M. De Haan & M. R. Gunnar (Eds.), *The handbook of developmental social neuroscience* (pp. 419–434). New York, NY: Guilford Press.

Blair, R. J., Mitchell, D. G., & Blair, K. (2005). *The Psychopath: Emotion and the brain.* Malden, MA: Blackwell.

Boccardi, M., Frisoni, G. B., Hare, R. D., Cavedo, E., Najt, P., Pievani, M., . . . Tiihonen, J. (2011). Cortex and amygdala morphology in psychopathy. *Psychiatry Research: Neuroimaging, 193,* 85–92. http://dx.doi.org/10.1016/j.pscychresns.2010.12.013

Brener, N. D., Billy, J. O., & Grady, W. R. (2003). Assessment of factors affecting the validity of self-reported health-risk behavior among adolescents: Evidence from the scientific literature. *Journal of Adolescent Health, 33,* 436–457. http://dx.doi.org/10.1016/S1054-139X(03)00052-1

Burgdorf, J., & Panksepp, J. (2006). The neurobiology of positive emotions. *Neuroscience and Biobehavioral Reviews, 30,* 173–187. http://dx.doi.org/10.1016/j.neubiorev.2005.06.001

Craig, M. C., Catani, M., Deeley, Q., Latham, R., Daly, E., Kanaan, R., . . . Murphy, D. G. (2009). Altered connections on the road to psychopathy. *Molecular Psychiatry, 14,* 946–953. http://dx.doi.org/10.1038/mp.2009.40

Crespi, B., & Badcock, C. (2008). Psychosis and autism as diametrical disorders of the social brain. *Behavioral and Brain Sciences, 31,* 241–261. http://dx.doi.org/10.1017/S0140525X08004214

Cuniff, M. M. (2010, December 16). Teen kills dogs after fight with mother, police say. *The Olympian.* Retrieved from http://www.theolympian.com/news/article25273582.html

Dadds, M. R., Whiting, C., & Hawes, D. J. (2006). Associations among cruelty to animals, family conflict, and psychopathic traits in childhood. *Journal of Interpersonal Violence, 21,* 411–429. http://dx.doi.org/10.1177/0886260505283341

Davis, M. (1992). The role of the amygdala in fear and anxiety. *Annual Review of Neuroscience, 15,* 353–375. http://dx.doi.org/10.1146/annurev.ne.15.030192.002033

Davis, M., & Whalen, P. J. (2001). The amygdala: Vigilance and emotion. *Molecular Psychiatry, 6,* 13–34. http://dx.doi.org/10.1038/sj.mp.4000812

de Waal, F. B. (2008). Putting the altruism back into altruism: The evolution of empathy. *Annual Review of Psychology, 59*, 279–300. http://dx.doi.org/10.1146/annurev.psych.59.103006.093625

Decety, J. (2015). The neural pathways, development and functions of empathy. *Current Opinion in Behavioral Sciences, 3*, 1–6.

Desimone, R., & Duncan, J. (1995). Neural mechanisms of selective visual attention. *Annual Review of Neuroscience, 18*, 193–222. http://dx.doi.org/10.1146/annurev.ne.18.030195.001205

Ermer, E., Cope, L. M., Nyalakanti, P. K., Calhoun, V. D., & Kiehl, K. A. (2012). Aberrant paralimbic gray matter in criminal psychopathy. *Journal of Abnormal Psychology, 121*, 649–658.

Glenn, A. L., Raine, A., & Schug, R. A. (2009). The neural correlates of moral decision-making in psychopathy. *Molecular Psychiatry, 14*, 5–6. http://dx.doi.org/10.1038/mp.2008.104

Goddard, G. V. (1964). Functions of the amgydala. *Psychological Bulletin, 62*, 89–109. http://dx.doi.org/10.1037/h0044853

Hala, S., Chandler, M., & Fritz, A. S. (1991). Fledgling theories of mind: Deception as a marker of three-year-olds' understanding of false belief. *Child Development, 62*, 83–97. http://dx.doi.org/10.2307/1130706

Hammer, J. L., & Marsh, A. A. (2015). Why do fearful facial expressions elicit behavioral approach? Evidence from a combined approach-avoidance implicit association test. *Emotion, 15*, 223–231. http://dx.doi.org/10.1037/emo0000054

Hare, R. D. (1991). *The Hare Psychopathy Checklist–Revised.* Toronto, Ontario, Canada: Multi-Health Systems.

Harenski, C. L., Harenski, K. A., Shane, M. S., & Kiehl, K. A. (2010). Aberrant neural processing of moral violations in criminal psychopaths. *Journal of Abnormal Psychology, 119*, 863–874. http://dx.doi.org/10.1037/a0020979

Izard, C. E. (2007). Basic emotions, natural kinds, emotion schemas, and a new paradigm. *Perspectives on Psychological Science, 2*, 260–280. http://dx.doi.org/10.1111/j.1745-6916.2007.00044.x

Jones, A. P., Laurens, K. R., Herba, C. M., Barker, G. J., & Viding, E. (2009). Amygdala hypoactivity to fearful faces in boys with conduct problems and callous-unemotional traits. *The American Journal of Psychiatry, 166*, 95–102. http://dx.doi.org/10.1176/appi.ajp.2008.07071050

Krebs, D. (1975). Empathy and altruism. *Journal of Personality and Social Psychology, 32*, 1134–1146. http://dx.doi.org/10.1037/0022-3514.32.6.1134

LeDoux, J. (2003). The emotional brain, fear, and the amygdala. *Cellular and Molecular Neurobiology, 23*(4-5), 727–738. http://dx.doi.org/10.1023/A:1025048802629

Lingle, S., & Riede, T. (2014). Deer mothers are sensitive to infant distress vocalizations of diverse mammalian species. *American Naturalist, 184*, 510–522. http://dx.doi.org/10.1086/677677

Lorenz, K. (1966). *On aggression*. London, England: Methuen.

Lozier, L. M., Cardinale, E. M., VanMeter, J. W., & Marsh, A. A. (2014). Mediation of the relationship between callous-unemotional traits and proactive aggression by amygdala response to fear among children with conduct problems. *JAMA Psychiatry, 71*, 627–636. http://dx.doi.org/10.1001/jamapsychiatry.2013.4540

Luk, E. S., Staiger, P. K., Wong, L., & Mathai, J. (1999). Children who are cruel to animals: A revisit. *Australian and New Zealand Journal of Psychiatry, 33*, 29–36. http://dx.doi.org/10.1046/j.1440-1614.1999.00528.x

Lykken, D. T. (1995). *The antisocial personalities*. Hillsdale, NJ: Erlbaum.

Maren, S. (2001). Neurobiology of Pavlovian fear conditioning. *Annual Review of Neuroscience, 24*, 897–931. http://dx.doi.org/10.1146/annurev.neuro.24.1.897

Marsh, A. A., Adams, R. B., Jr., & Kleck, R. E. (2005). Why do fear and anger look the way they do? Form and social function in facial expressions. *Personality and Social Psychology Bulletin, 31*, 73–86. http://dx.doi.org/10.1177/0146167204271306

Marsh, A. A., & Ambady, N. (2007). The influence of the fear facial expression on prosocial responding. *Cognition and Emotion, 21*, 225–247. http://dx.doi.org/10.1080/02699930600652234

Marsh, A. A., Ambady, N., & Kleck, R. E. (2005). The effects of fear and anger facial expressions on approach- and avoidance-related behaviors. *Emotion, 5*, 119–124. http://dx.doi.org/10.1037/1528-3542.5.1.119

Marsh, A. A., & Blair, R. J. (2008). Deficits in facial affect recognition among antisocial populations: A meta-analysis. *Neuroscience and Biobehavioral Reviews, 32*, 454–465. http://dx.doi.org/10.1016/j.neubiorev.2007.08.003

Marsh, A. A., & Cardinale, E. M. (2012). Psychopathy and fear: Specific impairments in judging behaviors that frighten others. *Emotion, 12*, 892–898. Advance online publication. http://dx.doi.org/10.1037/a0026260

Marsh, A. A., Finger, E. C., Fowler, K. A., Jurkowitz, I. T., Schechter, J. C., Yu, H. H., . . . Blair, R. J. (2011). Reduced amygdala-orbitofrontal connectivity during moral judgments in youths with disruptive behavior disorders and psychopathic traits. *Psychiatry Research: Neuroimaging, 194*, 279–286. http://dx.doi.org/10.1016/j.pscychresns.2011.07.008

Marsh, A. A., Finger, E. C., Mitchell, D. G. V., Reid, M. E., Sims, C., Kosson, D. S., . . . Blair, R. J. R. (2008). Reduced amygdala response to fearful expressions in children and adolescents with callous-unemotional traits and disruptive behavior disorders. *The American Journal of Psychiatry, 165*, 712–720. http://dx.doi.org/10.1176/appi.ajp.2007.07071145

Marsh, A. A., Kozak, M. N., & Ambady, N. (2007). Accurate identification of fear facial expressions predicts prosocial behavior. *Emotion, 7*, 239–251. http://dx.doi.org/10.1037/1528-3542.7.2.239

Motzkin, J. C., Newman, J. P., Kiehl, K. A., & Koenigs, M. (2011). Reduced prefrontal connectivity in psychopathy. *The Journal of Neuroscience, 31*, 17348–17357. http://dx.doi.org/10.1523/JNEUROSCI.4215-11.2011

Muñoz, L. C. (2009). Callous-unemotional traits are related to combined defi-cits in recognizing afraid faces and body poses. *Journal of the American Acad-emy of Child & Adolescent Psychiatry, 48,* 554–562. http://dx.doi.org/10.1097/CHI.0b013e31819c2419

Newman, J. P. (1998). Psychopathic behaviour: An information processing perspec-tive. In D. J. Cooke, A. E. Forth, & R. D. Hare (Eds.), *Psychopathy: Theory, research, and implications for society* (pp. 81–104). Dordrecht, The Netherlands: Kluwer Academic. http://dx.doi.org/10.1007/978-94-011-3965-6_5

Nichols, S. (2001). Mindreading and the cognitive architecture underlying altru-istic motivation. *Mind & Language, 16,* 425–455. http://dx.doi.org/10.1111/1468-0017.00178

Raine, A. (2008). From genes to brain to antisocial behavior. *Current Directions in Psychological Science, 17,* 323–328. http://dx.doi.org/10.1111/j.1467-8721.2008.00599.x

Rilling, J. K., Glenn, A. L., Jairam, M. R., Pagnoni, G., Goldsmith, D. R., Elfenbein, H. A., & Lilienfeld, S. O. (2007). Neural correlates of social cooperation and non-cooperation as a function of psychopathy. *Biological Psychiatry, 61,* 1260–1271. http://dx.doi.org/10.1016/j.biopsych.2006.07.021

Schenkel, R. (1967). Submission: Its features in the wolf and dog. *American Zoologist, 7,* 319–329.

Scott, S. K., Young, A. W., Calder, A. J., Hellawell, D. J., Aggleton, J. P., & Johnsons, M. (1997). Impaired auditory recognition of fear and anger fol-lowing bilateral amygdala lesions. *Nature, 385,* 254–257. http://dx.doi.org/10.1038/385254a0

Skeem, J. L., Polaschek, D. L. L., Patrick, C. J., & Lilienfeld, S. O. (2011). Psycho-pathic personality: Bridging the gap between scientific evidence and public policy. *Psychological Science in the Public Interest, 12,* 95–162. http://dx.doi.org/10.1177/1529100611426706

Smith, J. M., & Price, G. R. (1973). The logic of animal conflict. *Nature, 246,* 15–18. http://dx.doi.org/10.1038/246015a0

Sprengelmeyer, R., Young, A. W., Schroeder, U., Grossenbacher, P. G., Federlein, J., Büttner, T., & Przuntek, H. (1999). Knowing no fear. *Proceedings of the Royal Society of London, Series B: Biological Sciences, 266*(1437), 2451–2456. http://dx.doi.org/10.1098/rspb.1999.0945

Sylvers, P. D., Brennan, P. A., & Lilienfeld, S. O. (2011). Psychopathic traits and preattentive threat processing in children: A novel test of the fearlessness hypothesis. *Psychological Science, 22,* 1280–1287. http://dx.doi.org/10.1177/0956797611420730

White, S. F., Marsh, A. A., Fowler, K. A., Schechter, J. C., Adalio, C., Pope, K., . . . Blair, R. J. (2012). Reduced amygdala response in youths with disruptive behavior disorders and psychopathic traits: Decreased emotional response versus increased top-down attention to nonemotional features. *The American*

Journal of Psychiatry, 169, 750–758. http://dx.doi.org/10.1176/appi.ajp.
2012.11081270

Yang, Y., Raine, A., Narr, K. L., Colletti, P., & Toga, A. W. (2009). Localization of
deformations within the amygdala in individuals with psychopathy. *Archives of
General Psychiatry, 66,* 986–994. http://dx.doi.org/10.1001/archgenpsychiatry.
2009.110

Zebrowitz, L. A., & Montepare, J. M. (1992). Impressions of babyfaced individuals across the life span. *Developmental Psychology, 28,* 1143–1152. http://dx.doi.
org/10.1037/0012-1649.28.6.1143

INTEGRATIVE COMMENTARY II: SHARED NEUROBIOLOGICAL MECHANISMS AND SOCIAL INTERACTIONS IN HUMAN–ANIMAL INTERACTION

LISA S. FREUND

A major goal of this volume is to bridge the discipline of social neuro-science with research involving human–animal interactions (HAI). Using a social neuroscience approach to HAI research will help identify the neuro-biological mechanisms underlying the effects of HAI and offer new para-digms for investigating those effects. Additionally, a focus on HAI, as a social activity for both the humans and nonhuman animals involved, has the promise of opening new areas of understanding of human social func-tioning within the domain of social neuroscience. The chapters in this part, Applying Neuroscience to Human–Animal Interaction, provide a strong rationale for making the effort to build a bridge between social neuroscience and HAI.

http://dx.doi.org/10.1037/14856-010
The Social Neuroscience of Human–Animal Interaction, L. S. Freund, S. McCune, L. Esposito, N. R. Gee, and P. McCardle (Editors)

NEUROENDOCRINE AND AUTONOMIC
MECHANISMS RELATED TO HAI

Carter and Porges (Chapter 4) clearly lay out what is known about the neuroendocrine and autonomic mechanisms underlying biological conse-quences of positive social interactions. Most of what we understand scientifi-cally about brain mechanisms (neuroendocrine) and autonomic physiological mechanisms associated with positive social interactions comes from research done with mammals, primarily rats, mice, and voles (murids). Specifically, this knowledge has come through the study of murid maternal behaviors. It is from this research that the hormones oxytocin and vasopressin have been identified as contributing components in brain mechanisms underlying social behaviors (e.g., Ahern & Young, 2009; Bales, Boone, Epperson, Hoffman, & Carter, 2011). Oxytocin is associated with sociality and positive social and nurturing behaviors, and vasopressin is associated with animal pair-bonding and paternal defense of offspring (Carter et al., 1997).

Other neural contributors to social behaviors are neurotransmitters, such as dopamine (on which the effects of oxytocin and vasopressin depend), which is associated with the sense of reward or satisfaction from the devel-opment of positive social bonding (Love, 2014). The coordinated systems underlying social behaviors are complex and interdependent: Positive social interaction behaviors depend on a quiescent, steady-state autonomic nervous system, and the autonomic nervous system depends in turn on the neural mechanisms and functions of oxytocin and vasopressin, which maintain posi-tive social interactions, which in turn increase the activity of oxytocin and vasopressin brain receptors, and so on.

The major point of the Carter and Porges chapter is that humans have "built-in," phylogenetically conserved DNA sequences leading to neural sys-tems primed for social interaction and that we share these DNA sequences with other mammals. We are primed at birth to engage with others. We are born with the biological architecture to develop social awareness and engage socially. As attachment, bonding, and broader social interaction experiences develop over time, social behaviors mature and the social brain becomes more sophisticated and able to detect social nuance. Why humans (and many other mammals) evolved with a social brain is not well understood, but it has been proposed that a need for socially monogamous pair-bonds developed because of the labor-intensive, protracted task of raising children (Rilling & Young, 2014). Pair-bonds and maternal/paternal care of children require social coordination between partners or parents, and this requires a more developed social brain. Indeed, the ability to process even subtle social sig-nals between individuals (both sending and receiving) is likely necessary for human (and nonhuman animal) survival. Furthermore, the human brain is wired such that we experience reward during mutual social interactions and feel negative emotions if socially rejected. We contend, and others have proposed (Carter & Porges, 2012; Decety, 2011; Rilling & Young, 2014), that this social wiring of the human brain developed through phylogenetically conserved neural circuits found in lower subclasses of mammals supporting more rudimentary social behaviors.

Developmental research, both human and comparative, has shown that neonates are born with several capabilities, including an attention bias to the face (Pascalis, de Schonen, Morton, Deruelle, & Fabre-Grenet, 1995; Paukner, Huntsberry, & Suomi, 2010), and can imitate facial gestures (Meltzoff & Moore, 1983; Paukner, Simpson, Ferrari, Mrozek, & Suomi, 2014; Striano, Henning, & Stahl, 2005), all in support of attachment and bonding. Human infants quickly learn reciprocal interactions in a social context. Infants also show early capabilities in other domains, such as being able to predict physical properties of objects they have never touched (Wang, Baillargeon, & Paterson, 2005). Infants and other nonhuman animals even show very early evidence of a mental approximate number system. For example, Feigenson, Dehaene, and Spelke (2004) found that 6-month-old infants are sensitive to differences between quantities of 8 and 16. Zhang et al. (2014) reported that the human neonate brain can distinguish between emotional voices of fear and anger. In fact, many behavioral neuroscientists have discussed the idea of various human "core competencies" for which the human (and in many cases, any mammal) brain is selectively built. So it is not surprising that we are also "wired" for social interaction, even cross-species social interaction. It is a human and nonhuman mammal core competency.

HAI—ATTACHMENT, INTERSUBJECTIVITY, AND EMPATHY

As Beetz and Bales (Chapter 5) explain, development of the human social brain originates within the parent/caregiver–infant bond or attachment and is supported by the oxytocin system, whose functioning in the child is facilitated by the caregiver through patterns of parental care, which is further supported by the oxytocin system in the parent. In fact, disrupting brain oxytocin signaling, pharmacologically or genetically, does disrupt maternal behavior in mice and can be restored by oxytocin injection. In human parents, oxytocin from their plasma is positively correlated with affectionate contact (Apter-Levi, Zagoory-Sharon, & Feldman, 2014; Feldman et al., 2012). To further support the evidence that oxytocin is important for both parent and child during social interactions, Weisman, Zagoory-Sharon, and Feldman (2012) administered either oxytocin or a placebo intranasally to fathers of 5-month-old infants and watched their interactions. There were increased salivary oxytocin and positive parenting behaviors, such as touch and social reciprocity, in the fathers receiving the intranasal oxytocin when compared with fathers in the placebo group. In parallel, the infants of the fathers given oxytocin showed increased peripheral oxytocin and displayed more social gaze and exploratory behavior than the infants of fathers given a placebo. Thus, there were parallel biological and behavioral positive effects on the child when the child's father was administered oxytocin, without direct hormonal manipulation to the infant (Weisman et al., 2012). The message is that the young infant is born with the neurobiological blueprint to support social development and that development is facilitated further by the nurturing behaviors of caregivers, which are supported by the caregivers' own neurobiological systems.

Beetz and Bales (Chapter 5) also review the positive effects of HAI on humans, including the *social catalyst effect* of animals (wherein the presence of animals increases social interactions among humans of all ages), facilitation of positive mood, buffering of stress systems, and contribution to general health benefits. The authors bring together several biobehavioral theories relevant to HAI, including polyvagal theory, attachment theory, and classical conditioning and distraction effects. More specifically, they suggest that oxytocin may play a role in human–animal bonding in processes similar to those associated with maternal–infant bonding and parental behavior.

It is not far-fetched to propose that human relationships with non-human companion animals can meet the criteria for human-to-human attachment as defined by the Ainsworth (1991) attachment theory. The Ainsworth criteria for attachment are that the social bond is persistent, involves a particular person (or nonhuman animal), is emotionally significant, involves proximity or contact sought with the subject of the social bond, and invokes feelings of sadness or distress at involuntary separation. Perhaps most significantly, one seeks security and comfort from those with whom he or she is attached. As Beetz and Bales point out, the opportunity for caregiving behavior is unmistakably present with companion animals, such as dogs, cats, gerbils, or rabbits. What we do not have is a well-established, scientific evidence base for whether the bond between humans and companion animals can activate the same neural systems in humans as those found for human parents and their offspring. We also do not have solid evidence of whether nonhuman animals themselves meet the criteria for attachment with a human, or if the same neural systems as those in humans support the establishment and maintenance of that attachment bond.

Classical conditioning is not often mentioned in connection with social neuroscience, but Beetz and Bales make a strong case. They suggest that a classical conditioning hypothesis could be used to investigate the positive effects of HAI. For example, a classical conditioning approach would state that the human associates the animal with relaxation, and the animal becomes a conditioned stimulus that triggers a physiological response.

Beetz and Bales also ask that researchers investigating effects in animal–assisted therapy (AAT) settings look at the possibility of the animal acting as a distractor from anxiety, pain, or shyness experienced in therapy sessions. Determining whether AAT effects result from distraction, increased motivation, or activation of biobehavioral systems supporting greater engagement during a session is an area of HAI that requires careful parsing through well-controlled experiments or randomized controlled trials.

Beetz and Bales make an excellent point about including a focus on the nonhuman animal's side of social interaction. As has been iterated in this commentary, there are shared neurobiological systems in humans and non-human animals. Thus, there is reason to predict that the positive effects of social interaction between humans and animals can translate to the animal.

Social interactions depend on what social neuroscientists refer to as *intersubjectivity* (how a person understands and relates to another), considered a main element of human sociality (Rochat, Passos-Ferreira, & Salem, 2009). At its most rudimentary, we are talking about imitation, present at birth in humans. With typical development, an infant will show reciprocity, engage in joint attention with another toward some object or activity, and distinguish his or her "self" from others. Eventually, the young child develops an understanding of the intentions or internal state of another, and this "sensitivity" to others can continue to develop into adulthood. A minimum degree of intersubjectivity in terms of understanding the intentions or internal state of another is necessary for a human to express empathy toward others. Understanding how and when empathy is evoked and expressed toward or from another person, a nonhuman animal, or even a cartoon character is an important domain of research for understanding of human social behavior. Lozier, Brethel-Haurwitz, and Marsh (Chapter 7) explore the concept of empathy, what it comprises, what it is not (psychopathy), its neural basis, and a range of methods social neuroscientists use to assess empathy.

Similar to the earlier discussion of phylogenetically conserved neural supports of social behavior in mammals (human and nonhuman), Decety (2011) pointed out that empathy is built on "deep evolutionary, biochemical, and neurological underpinnings" (p. 35) and the "phylogenetically conserved neural circuits that support social behavior" (Decety & Michalska, 2012, p. 168). Empathy relies on basic core mechanisms associated with affective communication (which can be nonverbal or verbal), social attachment, and parental care. Similar systems in humans and nonhuman animals that regulate parental behavior and emotional processing, when combined with the higher level cognitive abilities of the human, produce the flexible and generalized forms of bonding and nurturing care found among humans. According to Decety, this explains why humans not only care for offspring, but can also care for the welfare of strangers and even bond with a nonhuman animal.

Nonhuman animals, too, even rodents, show evidence of what at least some neuroscientists are calling "empathy." In one study (Bartal, Decety, & Mason, 2011), free rats intentionally and quickly opened a restraint to free a distressed, restrained cagemate. When liberating the cagemate, the rats typically shared a chocolate treat with the distressed rat. We do not know whether the free rats were actually experiencing what we know as empathy or just the desire for social contact (Silberberg et al., 2014), but their behaviors were certainly analogous to behaviors we call empathetic in humans.

THE NEUROSCIENCE OF EMPATHY

Social neuroscientists have begun providing evidence of the neurological basis of empathy in humans. As Lozier, Brethel-Haurwitz, and Marsh point out, areas of the brain that appear to be involved with processing information and emotions related to empathy are the amygdala (involved with emotional coding of social signals) and its connections to hippocampus, orbitofrontal and ventromedial cortex, and striatum (which mediate processes of central emotion response, learning, and memory). Assessing human brain activation through functional magnetic resonance imaging (fMRI) while one is exposed to secondhand experiences of pain by observing other individuals in pain is one way social neuroscientists have assessed empathy and consistently identified the brain areas associated with empathy (see also Jackson, Brunet, Meltzoff, & Decety, 2006; Lamm, Batson, & Decety, 2007). When exposed to images of people getting hurt, participants undergoing an fMRI show more activity in the same neural circuits that process the firsthand experiences of pain (Yang, Decety, Lee, Chen, & Cheng, 2009). Some scientists have suggested that this neural phenomenon is reflecting the activation of certain neurons, called mirror neurons, which are nerve cells that fire both when a person performs an action and when he or she sees that action performed by others. Activated mirror neurons have even been seen in very young human infants (Marshall & Meltzoff, 2014) and nonhuman primates (both infant and adult; Paukner et al., 2014), although the specific neurons have only been identified when mirroring motor movements; mirror neurons specific to emotions or empathy have not yet been pinpointed.

The Lozier, Brethel-Haurwitz, and Marsh chapter develops a fuller understanding of empathy by exploring manifestations of the lack of empathy, which involve shallow emotional display, difficulty forming close bonds with others, or cruelty toward others. Lack of emotional bonds or the inability to feel the pain of another appears to involve deficits or differences in the functioning of neural areas and circuits associated with some aspects of empa-thy, such as lower activation of amygdala functioning or reduced functional connectivity between the amygdala and frontal areas of the brain associated with emotional control. The disturbing occurrences of animal cruelty, the extreme negative aspect of HAI, may well be associated with dysfunction in these key brain circuits, but this putative association has not been scientifi-cally established and remains an open research question.

HAI AS SOCIAL SUPPORT FOR EMOTIONAL SELF-REGULATION

Social bonding and positive behaviors can attenuate stress-related activities; support important developmental cognitive, behavioral, neural, and other biological processes; and maintain the social environment through prosocial expressions such as empathy. Each of the chapters in this part has pointed to the brain's architecture for social interaction, with Brown and Coan (Chapter 6) stating that "the brain is acutely social." The brain is primed for supporting social interactions, and it relies on social supports for healthy functioning. Socially adaptive human beings develop self-regulation of behaviors, cognition, and emotions. As Brown and Coan point out, such self-regulation requires energy for functions in specific brain structures such as the prefrontal cortex (PFC). Humans have developed a large PFC com-pared with other animals, but there are limits to the brain energy available for self-regulatory processes. Brown and Coan contend that these limits in neural energy for self-regulation can be mitigated by social supports, and research supports this idea. When exposed to a perceived threat, functional neuroim-aging has shown that human participants' PFCs were less active (less energy expended) during the context of supportive social interaction than when no social support was available (Coan, Beckes, & Allen, 2013; Coan, Schaefer, & Davidson, 2006). As Brown and Coan put it, social support is an efficient and energy cost-effective means of controlling (self-regulating) emotional responses and stress reactions.

If one accepts that HAI involves a type of meaningful, positive social interaction for humans, then one can hypothesize that companion animals or other positive nonhuman animal interactions can assist in the human's self-regulation of emotion and stress responses. That is, by reducing stress or the perception of threat to one's brain or physical energy resources, a companion animal can provide a human with a sense of increased internal resources for dealing with stressful life events, everyday hassles, or perceived threats to well-being. Whether this effect of HAI would show as reduced PFC activation as in the studies by Coan and colleagues (Coan et al., 2006, 2013) or as another neural signature indicating perceived safety associated with an attachment figure (Eisenberger et al., 2011) is an area ripe for further research.

NEUROSCIENCE TECHNOLOGIES AND FUTURE RESEARCH DIRECTIONS

Throughout the chapters in this part, it is apparent that social neuroscience research relies heavily on technologies for looking at brain neurochemistry, brain structures and function, and physiological responses associated with the central nervous system. For example, electroencephalography and fMRI are used to determine levels of brain activation associated with exposure to a visual or auditory stimulus or a cognitive task. The use of fMRI related to HAI is shown in a recent study where human brain activity was measured in response to positive or negatively valenced vocalizations generated by cats (a familiar animal), rhesus monkeys (a less familiar animal), and humans and compared with nonbiological sounds (Belin et al., 2008). Humans can differentiate between negative and positive vocalizations for both human and nonhuman animals, indicating a common human brain response to mammal emotional expression regardless of species. In another fascinating study, a Hungarian team trained dogs to lie motionless in MRI machines while listing to vocal emotional cues and nonvocal sounds. The fMRI results for the dogs and for humans listening to the same sounds indicated that dogs' and humans' brains show striking similarities in the way both species process emotionally loaded sounds (Andics, Gácsi, Faragó, Kis, & Miklósi, 2014).

Another technology that is starting to be used more often in social neuroscience research is that of eye tracking for understanding the control of attention and degree of social orienting. Using an eye-tracking device that allows freedom of head movement, pet dogs watched video presentations of a human actor turning toward one of two objects, and their eye-gaze patterns were recorded (Téglás, Gergely, Kupán, Miklósi, & Topál, 2012). Dogs were more likely to follow where the human looked if the human preceded turning with an expression of communicative intent (direct gaze at the dog). Thus, this research shows both that eye-tracking techniques can be used for studying dogs' social skills and that dogs respond to expressions of communicative intent from humans. The research also leads one to wonder whether dual eye tracking of human and companion pet eye gazes during interactions might yield interesting information about human and nonhuman animal social reciprocity, attachment, intersubjectivity, or empathy during HAI encounters.

Given the degree of overlap among the neural systems supporting human and nonhuman animal positive social behaviors, there is clearly a role for HAI to play in social neuroscience research and a role for the methods of social neuroscience to play for HAI researchers. Research questions that can be explored through the context of this overlap could include (but are certainly not limited to) the following: How are HAIs the same or different from human–human social interactions and how does that increase our understanding of human social functioning? Do HAIs impact neural plasticity? What is the reciprocal action of oxytocin between a human and a nonhuman animal during HAI? What is the impact of social touch during HAI and what are the underlying mechanisms? Is a human's joint attention with a companion animal the same as with another human? Do humans show mirror neuron function toward nonhuman animals? Does HAI facilitate emotion regulation in young children when this ability is not yet mature? How does HAI function differently during different periods in the human life span and does that inform us about social development? Tapping into a nonverbal yet reciprocal social interaction may also be a particularly fruitful approach for understanding the human ability to pick up very subtle social nonverbal cues or prosody in language or in the sounds emitted by nonhuman animals. This could also be important, especially when intervening to improve human communicative social deficits associated with human developmental disabilities or degenerative decline in advanced age. The bottom line is that the science literature, measures, and approaches of social neuroscience are there for HAI researchers to tap into, and the opportunities for assessing broader concepts of social interaction, attachment, intersubjectivity, empathy, and social regulation of emotional responses are available for social neuroscientists through HAIs.

REFERENCES

Ahern, T. A., & Young, L. J. (2009). The impact of early life family structure on adult social attachment, alloparental behavior, and the neuropeptide systems regulating affiliative behaviors in the monogamous prairie vole (*Microtus ochrogaste*). *Frontiers in Behavioral Neuroscience*, 3, 1–19. http://dx.doi.org/10.3389/neuro.08.017.2009

Ainsworth, M. D. S. (1991). *Attachment and other affectional bonds across the life cycle*. New York, NY: Routledge.

Andics, A., Gácsi, M., Faragó, T., Kis, A., & Miklósi, Á. (2014). Voice-sensitive regions in the dog and human brain are revealed by comparative fMRI. *Current Biology*, 24, 574–578. http://dx.doi.org/10.1016/j.cub.2014.01.058

Apter-Levi, Y., Zagoory-Sharon, O., & Feldman, R. (2014). Oxytocin and vasopressin support distinct configurations of social synchrony. *Brain Research*, 1580, 124–132. http://dx.doi.org/10.1016/j.brainres.2013.10.052

Bales, K. L., Boone, E., Epperson, P., Hoffman, G., & Carter, C. S. (2011). Are behavioral effects of early experience mediated by oxytocin? *Frontiers in Psychiatry, 2,* 1–12. http://dx.doi.org/10.3389/fpsyt.2011.00024

Bartal, I. B.-A., Decety, J., & Mason, P. (2011). Empathy and pro-social behavior in rats. *Science, 334,* 1427–1430. http://dx.doi.org/10.1126/science.1210789

Belin, P., Fecteau, S., Charest, I., Nicastro, N., Hauser, M. D., & Armony, J. L. (2008). Human cerebral response to animal affective vocalizations. *Proceedings of the Royal Society Series B, 275,* 473–481. http://dx.doi.org/10.1098/rspb.2007.1460

Carter, C. S., DeVries, A. C., Taymans, S. E., Roberts, R. L., Williams, J. R., & Getz, L. L. (1997). Peptides, steroids, and pair bonding. *Annals of the New York Academy of Sciences, 807,* 260–272. http://dx.doi.org/10.1111/j.1749-6632.1997.tb51925.x

Carter, C. S., & Porges, S. W. (2012). The biochemistry of love: An oxytocin hypothesis. *EMBO Reports, 14,* 12–16. http://dx.doi.org/10.1038/embor.2012.191

Coan, J. A., Beckes, L., & Allen, J. P. (2013). Childhood maternal support and social capital moderate the regulatory impact of social relationships in adulthood. *International Journal of Psychophysiology, 88,* 224–231. http://dx.doi.org/10.1016/j.ijpsycho.2013.04.006

Coan, J. A., Schaefer, H. S., & Davidson, R. J. (2006). Lending a hand: Social regulation of the neural response to threat. *Psychological Science, 17,* 1032–1039. http://dx.doi.org/10.1111/j.1467-9280.2006.01832.x

Decety, J. (2011). The neuroevolution of empathy. *Annals of the New York Academy of Sciences, 1231,* 35–45. http://dx.doi.org/10.1111/j.1749-6632.2011.06027.x

Decety, J., & Michalska, K. J. (2012). How children develop empathy: The contribution of developmental affective neuroscience. In J. Decety (Ed.), *Empathy: From bench to bedside* (pp. 167–191). Cambridge, MA: MIT Press.

Eisenberger, N. I., Master, S. L., Inagaki, T. K., Taylor, S. E., Shirinyan, D., Lieberman, M. D., & Naliboff, B. D. (2011). Attachment figures activate a safety signal-related neural region and reduce pain experience. *Proceedings of the National Academy of Sciences, USA, 108,* 11721–11726. http://dx.doi.org/10.1073/pnas.1108239108

Feigenson, L., Dehaene, S., & Spelke, E. (2004). Core systems of number. *Trends in Cognitive Sciences, 8,* 307–314. http://dx.doi.org/10.1016/j.tics.2004.05.002

Feldman, R., Zagoory-Sharon, O., Weisman, O., Schneiderman, I., Gordon, I., Maoz, R., . . . Ebstein, R. P. (2012). Sensitive parenting is associated with plasma oxytocin and polymorphisms in the OXTR and CD38 genes. *Biological Psychiatry, 72,* 175–181. http://dx.doi.org/10.1016/j.biopsych.2011.12.025

Jackson, P. L., Brunet, E., Meltzoff, A. N., & Decety, J. (2006). Empathy examined through the neural mechanisms involved in imagining how I feel versus how you feel pain. *Neuropsychologia, 44,* 752–761. http://dx.doi.org/10.1016/j.neuropsychologia.2005.07.015

Lamm, C., Batson, C. D., & Decety, J. (2007). The neural substrate of human empathy: Effects of perspective-taking and cognitive appraisal. *Journal of Cognitive Neuroscience*, *19*, 42–58. http://dx.doi.org/10.1162/jocn.2007.19.1.42

Love, T. M. (2014). Oxytocin, motivation and the role of dopamine. *Pharmacology, Biochemistry and Behavior*, *119*, 49–60. http://dx.doi.org/10.1016/j.pbb.2013.06.011

Marshall, P. J., & Meltzoff, A. N. (2014). Neural mirroring mechanisms and imitation in human infants. *Philosophical Transactions B*, *370*(1662), 1–11.

Meltzoff, A. N., & Moore, M. K. (1983). Newborn infants imitate adult facial gestures. *Child Development*, *54*, 702–709. http://dx.doi.org/10.2307/1130058

Pascalis, O., de Schonen, S., Morton, J., Deruelle, C., & Fabre-Grenet, M. (1995). Mother's face recognition by neonates: A replication and an extension. *Infant Behavior & Development*, *18*, 79–85. http://dx.doi.org/10.1016/0163-6383(95)90009-8

Paukner, A., Huntsberry, M. E., & Suomi, S. J. (2010). Visual discrimination of male and female faces by infant rhesus macaques. *Developmental Psychobiology*, *52*, 54–61.

Paukner, A., Simpson, E. A., Ferrari, P. F., Mrozek, T., & Suomi, S. J. (2014). Neonatal imitation predicts how infants engage with faces. *Developmental Science*, *17*, 833–840. http://dx.doi.org/10.1111/desc.12207

Rilling, J. K., & Young, L. J. (2014). The biology of mammalian parenting and its effect on offspring social development. *Science*, *345*, 771–776. http://dx.doi.org/10.1126/science.1252723

Rochat, P., Passos-Ferreira, C., & Salem, P. (2009). Three levels of intersubjectivity in early development. In A. Carassa, F. Morganti, & G. Riva (Eds.), *Enacting intersubjectivity: Paving the way for a dialogue between cognitive science, social science, social cognition and neuroscience* (pp. 173–190). Lugano, Switzerland: Universita Svizzera Italiana.

Silberberg, A., Allouch, C., Sandfort, S., Kearns, D., Karpel, H., & Slotnick, B. (2014). Desire for social contact, not empathy, may explain "rescue" behavior in rats. *Animal Cognition*, *17*, 609–618. http://dx.doi.org/10.1007/s10071-013-0692-1

Striano, T., Henning, A., & Stahl, D. (2005). Sensitivity to social contingencies between 1 and 3 months of age. *Developmental Science*, *8*, 509–518. http://dx.doi.org/10.1111/j.1467-7687.2005.00442.x

Téglás, E., Gergely, A., Kupán, K., Miklósi, Á., & Topál, J. (2012). Dogs' gaze following is tuned to human communicative signals. *Current Biology*, *22*, 209–212. http://dx.doi.org/10.1016/j.cub.2011.12.018

Wang, S. H., Baillargeon, R., & Paterson, S. (2005). Detecting continuity violations in infancy: A new account and new evidence from covering and tube events. *Cognition*, *95*, 129–173. http://dx.doi.org/10.1016/j.cognition.2002.11.001

Weisman, O., Zagoory-Sharon, O., & Feldman, R. (2012). Oxytocin administration to parent enhances infant physiological and behavioral readiness for social engagement. *Biological Psychiatry, 72,* 982–989. http://dx.doi.org/10.1016/j.biopsych.2012.06.011

Yang, C. Y., Decety, J., Lee, S., Chen, C., & Cheng, Y. (2009). Gender differences in the mu rhythm during empathy for pain: An electroencephalographic study. *Brain Research, 1251,* 176–184. http://dx.doi.org/10.1016/j.brainres.2008.11.062

Zhang, D., Liu, Y., Hou, X., Sun, G., Cheng, Y., & Luo, Y. (2014). Discrimination of fearful and angry emotional voices in sleeping human neonates: A study of the mismatch brain responses. *Frontiers in Behavioral Neuroscience, 8,* 422. http://dx.doi.org/10.3389/fnbeh.2014.00422

III

SCIENCE AND RESEARCH CONSIDERATIONS

8

GENETIC COMPONENTS OF COMPANION ANIMAL BEHAVIOR

PAUL JONES AND SANDRA McCUNE

The amazing amount of genetic variation present in dog and cat breeds is matched only by the behavioral characteristics observed in their interactions with their owners and other animals. More than 80 cat and 350 dog breeds are recognized by various clubs around the world, with each having "signature" or "stereotypical" behaviors that many owners use to make companion selections (see http://www.thekennelclub.org.uk/). Research into the genetic drivers of behavior is accelerating mainly due to progress in deoxyribonucleic acid (DNA) sequencing technologies and a fast decrease in sequencing cost. It is already possible to complete work in a single day that, 5 years ago, would have taken 3 years to complete. The research bottleneck, therefore, has moved to getting high-quality behavioral phenotypes and bioinformatics analysis of whole companion animal genomes.

A focus on defining good reference genomes (a representative example of a species' set of genes) for dogs and cats, funded by the National

http://dx.doi.org/10.1037/14856-011
The Social Neuroscience of Human–Animal Interaction, L. S. Freund, S. McCune, L. Esposito, N. R. Gee, and P. McCardle (Editors)

Institutes of Health, has facilitated genetic research on companion animals over the past 10 years. Genetic research is further advanced in dogs than in cats because of the earlier release of a Boxer reference genome and the larger range of known behaviors in dogs compared with cats. Cat research is now accelerating with the availability of an Abyssinian reference genome. Behavioral genetics has progressed from understanding domestication in dogs and cats, which produced many breed-cluster and breed-specific behaviors via genetic segregation, through to the study of the mechanisms underpinning behaviors. The pet behaviors under selection by humans since domestication have included pulling, hunting, protection, tracking, herding, companionship, and more recently lifestyle therapeutic assistance. These behaviors are more prevalent in some breeds, leading to the favoring of certain breeds for specific desired tasks.

This chapter summarizes the change in the genetic technology landscape, the types of genetic variation observed in pets, tools used to record behaviors, breed clusters in cats and dogs, breed-cluster and breed-specific behaviors and morphological and neurochemical genetic drivers of behavior. The next 10 years are likely to reveal detailed mechanisms of different animal behaviors that will benefit our understanding of pets and better enable us to tailor human–animal interaction (HAI).

EVOLUTION OF GENETIC TECHNOLOGIES

The past 10 years of genetic research in canids and felids have been limited by the cost of the technologies available to the researcher. The field has moved from one approach to many as those summarized in Table 8.1.

In the early 2000s, companion animal researchers were limited to candidate gene and microsatellite variation analyses. Candidate gene analyses at that time were mostly seeking to leverage discoveries made in human genetics to assess their relevance in companion animals. Microsatellites allowed researchers to analyze whole genomes, with the variation in the length of microsatellites on each chromosome being used to find associations to the phenotype of interest. Researchers investigating phenotypes driven by multiple genes using microsatellites developed quantitative loci trait maps of the chromosomes showing areas most likely to contain the genes driving the phenotype. These DNA techniques were complemented by research into ribonucleic acid (RNA) regulation in different tissues either by isolating RNA and creating stable versions of the tissue expression in the form of a complementary DNA (cDNA) library or by creating fluorescent complementary strands of DNA that were bound to gene

TABLE 8.1

Technologies Used to Study Genetic Variation in Companion Animals

Variation target	Technology	Notes
Candidate gene analysis	Sanger sequencing	Initially single gene analysis
Microsatellite analysis	Quantitative trait locus	Usually allows location of areas of interest to within a section of a chromosome
RNA expression	Expressed sequence tag libraries—cDNA sequencing	
	Gene expression arrays	
Single nucleotide polymorphisms	Whole genome association studies	
High throughput RNA expression	Transcriptomics	
Gene variation	Exomics	
Whole genome variation	Whole genome sequencing	Allows variations associated with phenotype to be found and interpreted in context of coding sequence and ENCODE regulatory sequences

Note. ENCODE = Encyclopedia of DNA Elements.

expression arrays. The cDNA library approach allowed detailed analysis of the sequence of individual RNA molecules, whereas the array approach revealed the level of a gene expression in the tissue of interest to be defined. Both of these approaches were correlated to the phenotype of interest to see if the sequence/structure or the expression level of individual RNA molecules was most likely to be driving the condition or behavior. The field of DNA analysis made a step change in capability in the mid-2000s when whole genome association studies became available, incorporating initially only a few hundred or thousand single nucleotide polymorphism (SNP) markers across the genome of interest (Grover & Sharma, 2014). In the late 2000s, a further advance in high throughput DNA sequencing by 454 Life Sciences and Illumina made the sequencing of DNA and RNA much more affordable. These technologies have enabled transcriptomics (generation of RNA expression profiles in selected tissues), exomics (sequencing of all the coding regions in a genome), and whole genome sequencing. The behavioral genetics researcher now has access to DNA and RNA tools that generate highly detailed information on every sample of interest.

GENETIC VARIATION IN CATS AND DOGS—
THE DRIVER OF BEHAVIORS

Companion animal genomes contain many different types of variation that can influence health, morphology, and behavior. The genetic variations studied to locate genes driving the phenotype of interest are listed in Table 8.2.

All of the genetic marker types mentioned previously have been studied in dogs and cats as drivers of health, morphology, or behaviors. Microsatellites

TABLE 8.2
Types of Genetic Variation Observed in Companion Animal Genomes

Variation type	Description	How the variation can impact phenotype
Microsatellite (MS)	Repeat of 2–6 nucleotides (e.g., a dinucleotide repeat of (CA)n where the repeat number n can be from 3 upwards)	Usually used to map locations rather than considered likely to be causal for the phenotype. Occasionally an MS may influence coding sequence or regulation of a gene.
Single nucleotide polymorphism (SNP)	A change of a single nucleotide where the commonly observed version is normal and the changed version referred to as the variant or mutation	SNPs can impact coding sequence, leading to a different protein sequence, or terminate the protein sequence early. SNPs can also impact regulatory regions (e.g., splice sites, repressor binding sites, transcription start sites).
Deletion	A deletion of 1 or more nucleotides in a contiguous stretch of DNA	Loss of a single to thousands or millions of nucleotides in a gene or regulatory regions often leads to phenotype change.
Insertion	Presence of 1 or more nucleotides at a location where the reference genome does not have the extra nucleotide(s)	Extra nucleotide(s) can act on coding or regulatory aspects of the animal phenotype.
Copy number variant	More than the usual number of copies of a gene	An increase in gene copy number can lead to higher levels of expression of the gene impacting the phenotype.
Retrotransposon	A mobile genetic element as in virus sequences that have taken up temporary or permanent residence within the companion animal genome	The integration site of the mobile element can impact coding or regulatory regions of the gene. Retrotransposons often have repeat elements like MSs that can continue to change in size by slippage mutation.

have been used in cats and dogs for unique identification of individuals, parentage testing, and linkage studies in large families where many of the pets have a quantified phenotype. Interestingly, the variation in microsatellites initially was shown to be higher in dogs than humans, implying that DNA variation across breeds may be wider than across humans, presumably because the rate of introduction of variation generation by generation may be higher in dogs (Fondon & Garner, 2004). Cats, conversely, do not seem to have unusually high microsatellite variation.

SNPs have been the focus of most genetic studies in companion animals for the past decade and therefore have been linked to more behaviors than other types of genetic variation. In dogs, SNPs have been found throughout the neuronal signaling and reception genes (Arata, Ogata, Shimozuru, Takeuchi, & Mori, 2008; Hejjas et al., 2009; Hejjas, Vas, Kubinyi, et al., 2007; Hejjas, Vas, Topal, et al., 2007; Kubinyi et al., 2012; Takeuchi et al., 2005, 2009) and are discussed later. Other than this study, cat SNP research has focused on morphology and health (e.g., coat color; Imes, Geary, Grahn, & Lyons, 2006; Ishida et al., 2006). Deletions, where a part of the chromosome or DNA sequence is missing, are classic drivers of variation impacting DNA regulatory elements or coding sequences. For example, insights into development and neurology are made possible by studies such as the 15.7 kb deletion in the *brevican* gene associated with episodic falling induced by exertion in Cavalier King Charles spaniels (Forman et al., 2012). In cats, the only behavioral variation linked to a deletion is the 247 coding sequence variant of *Tas1r2* gene that leads to cats not liking sweet foods, as the sugar receptor does not form as a heteromer protein (Li et al., 2005). This sugar sensor is also nonfunctional in other *felidae*, but it is not known whether it led to obligate carnivory, on the basis of savory meats effectively driving feeding behavior, or whether carnivory occurred first and the gene was lost as it was not important to survival.

The recent focus on copy number variation (CNV) as a driver of phenotypic variation is intriguing, with morphological phenotypes such as skin wrinkling in the Shar-Pei and the dorsal ridge in Rhodesian ridgebacks being driven by copy number (reviewed in Alvarez & Akey, 2012). CNV variation in dogs and cats has not yet been linked to behavioral phenotypes.

Retrotransposons are mobile genetic elements that may still retain their ability to replicate and move. Studies in humans with a retrotransposon called LINE-1 (long interspersed nuclear element), making up 20% of the human genome, have shown that this still mobile element "jumps" during neuronal development giving individual cells the opportunity to be more plastic through genetic alteration. In companion animals, retrotransposons have not yet been shown to jump during neuronal development. Initial studies have shown that the main SINE (short interspersed nuclear element) in dog (Can-SINE) occurs more than 40,000 times and was embedded in the genome before cats and dogs diverged. Cats and dogs share more than 500 SINEs, which have allowed phylogenetic studies of the species divergence (Schröder, Bleidorn, Hartmann, & Tiedemann, 2009). SINE evolution has continued since the incorporation into the cat and dog ancestral genomes, and some SINEs have been shown to be specific to domestic cats and dogs (Bentolila et al., 1999; Minnick, Stillwell, Heineman, & Stiegler, 1992; Pecon-Slattery, Pearks Wilkerson, Murphy, & O'Brien, 2004). In dogs, SINE elements have been found to be associated with or causal of changes in coat color (Clark, Tsai, Starr, Nowend, & Murphy, 2011; Dreger & Schmutz, 2011; Zeng et al., 2011), Neonatal Ataxia (in the coton de Tulear breed; Zeng et al., 2011), short leggedness (chondrodysplasia; Parker et al., 2009), mild hemophilia (in German wirehaired pointers; Brooks, Gu, Barnas, Ray, & Ray, 2003), and one behavioral phenotype conferring sleepiness/narcolepsy in some dogs (Lin et al., 1999).

Interestingly, dogs have been attributed with more morphological and behavioral diversity than any other land mammal (Ostrander, 2012), leading to a focus on the mechanisms that may underpin the ability of the species to be so developmentally plastic. Recently a recombination hotspot protein *PRDM9* has been shown to be disrupted in the dog (Axelsson et al., 2012), which may allow the proliferation of CNVs (Berglund et al., 2012). This mechanism is hypothesized by the authors and colleagues to be driving genome instability and chromosomal rearrangements in canids.

COMPANION ANIMAL BEHAVIORS

The behaviors of our pets have been analyzed in several ways to allow geneticists to search for genes that underpin the phenotypes of interest (Table 8.3).

TABLE 8.3
Methods Used to Record Behaviors in Companion Animal Studies

Method type	Description
Breed characteristics	Stereotype present in all members of the breed as defined by experts.
Survey tools	Individual characteristics recorded via diary or owner summary of the pet. Data can be rolled up to create a breed specific score.
Stimulus test— observer or video recording	Stimuli that provoke variation in behavioral responses that can be used to assess the robustness, curiosity, boldness of pets (often used by assistance groups such as guide dogs). Quantitative scales are usually used to assess the strength of response observed. This method allows geneticists to map behavioral variation against the variation present in each individual gene of the pet cohort.

The use of expert quantification of a behavior is useful to get rough genomic locations for the trait. Usually a large set of breeds (50 or more) is studied with whole genome arrays to associate the single behavioral score per breed to SNP variants. In one such study herding, pointing, boldness, and trainability loci were located within the dog genome (Jones et al., 2008).

Survey tools usually allow researchers to quickly access hundreds of samples for association studies. Many of these tools have been honed over the past few years, with the Canine Behavioral Assessment and Research Questionnaire (C-BARQ) being the gold standard reference design (Hsu & Serpell, 2003). C-BARQ seems to be robust across multiple geographies; similar behavioral profiles have been found in Japan, America, Taiwan, and the Netherlands (Hsu & Sun, 2010; Nagasawa et al., 2011; van den Berg, Schilder, de Vries, Leegwater, & van Oost, 2006). A feline version is currently under development at the time of writing.

Another tool for behavior quantification is challenge tests. Stimulus testing is used to challenge the pet with novel or preexperienced objects, sounds, or situations, and measures their response. The response can be recorded as a binary record (i.e., did or did not occur), or recordings can be broken down after the event into frequencies, durations, or sequences of behavior using an ethogram. The ethogram approach is the gold standard in this area: The order, time, and rate of the individual behavioral elements allow a much better understanding of behavioral sequencing and response to different stimuli (Berry et al., 2012; Webster, 2011).

ORIGINS OF BREED-SPECIFIC STEREOTYPES

Several breeds in both dogs and cats have characteristic behaviors, such as vocalization by Siamese and herding by collies. The origin of these behaviors can be understood by looking through a domestication lens to understand which were present in the progenitor animals and which have been developed by iterative breeding to better fit the intended purpose.

Dog Domestication as a Driver for Behavioral Abilities

Dogs, throughout history, have been selected for their abilities, both morphological and behavioral. Small terriers, for instance, have been used historically for hunting vermin and defending stored food because of their prey-seeking genetic programming. These small dogs have been present for more than 10,000 years alongside various human populations, emphasizing the important role that they have played. The grouping of dogs by their

TABLE 8.4
Dog Breed Groups Are Named for Their Behaviors

Breed group	Task	Behavior
Toy dogs	Earth dogs/rodent control	Hunt and chase behavior
Scent hounds	Hunting	Tracking
Herding	Livestock farming	Herding and guarding
Retrieving	Hunting	Retrieving

behaviors and roles (Table 8.4) is emphasized by genetic analysis of dogs that recapitulates the grouping by function/behavior.

Herding dogs, guard dogs, vermin hunters, and retrievers were developed into many different breeds while maintaining the behavioral function critical to HAI. Dog domestication and genetic selection by humans have led to breeds in these groups becoming specialized in key behaviors, with the interdependence of human and dog continually selecting for stronger interspecies dependencies. This long-term relationship seems to be driven by many independent behaviors and morphological features of dog breeds.

A two-stage hypothesis of domestication and socialization was presented by Udell, Dorey, and Wynne (2010), who hypothesized that some canids' acceptance of humans as social companions was the first stage and that dogs were subsequently conditioned to behaviors that are directed by the human partner (e.g., pointing at a bowl that is hiding a reward). In support of the second stage of the hypothesis, the ability of dogs to almost attribute the intent of a partner pointing at an object seems to be possible through strong associative learning, with food-based reward systems being a good reinforcement of the behavior (Elgier, Jakovcevic, Mustaca, & Bentosela, 2012; Feuerbacher & Wynne, 2012).

The ability of different breeds to respond to humans pointing at a target has been extended by comparing breeds that typically work closely with their human partner ("cooperative" herding and gun dogs) against breeds that work independent of their handler (e.g., guard dogs and earth dogs, chasing prey in burrows). These two breed clusters of cooperative versus independent breeds were compared in tasks involving pointing at a bowl containing food; the breeds described as cooperative performed better at the task than did independent breeds (Gácsi, McGreevy, Kara, & Miklósi, 2009).

HAI during training often relies on gaze acquisition before the indication of the task expected by the handler. An analysis of gaze acquisition and extinction among retrievers, German shepherd dogs, and poodles (Jakovcevic, Elgier, Mustaca, & Bentosela, 2010) revealed that all three breeds acquired gaze equally, but in retrievers the extinction of gaze occurred much later.

The reliance of gaze is also paramount in breed-specific motor patterns. Coppinger and Coppinger (2001) framed a very nice paradigm of behaviors linked to catching prey in the wolf, suggesting some steps of hunting behavior have adapted during the production of breeds to lead to very different behaviors. They suggested that certain breeds have a tendency not to naturally do some of the steps in the complete sequence of orient > eye > stalk > chase > grab > bite > kill > dissect behaviors seen in the wolf. For instance, herding breeds such as the collie have remarkable skills up to the chase phase, whereas livestock-guarding dogs (large sheepdogs and mastiffs) have lost all of the microbehaviors of prey capture, making them suitable to protect rather than stress, damage, or eat domesticated herds.

The method of breaking down natural behaviors into steps in this manner enables us to see where each microstep is an innate ability conferred by genetics and to consider that the sequencing of behaviors may be a reflection of the neurochemical links between the behavioral steps. This breakdown of a behavior, be it standing up, following a human point, or herding sheep, is useful for the geneticist both to interpret breed-specific behaviors and to understand mechanisms that might drive the microbehaviors.

The context of a stereotypical behavior of a group of breeds, a single breed, or an individual dog is therefore made possible by genetics driving the morphology, anatomy, and neurology of the companion animal.

Cat Domestication as a Driver for Behavioral Abilities

Domestication in cats is thought to have occurred where a proportion of the cat population that was more tolerant of humans cohabited with us and then became reliant on that niche. The Egyptians are thought to have recognized the usefulness of cats for vermin control (Messent, & Serpell, 1981), and there is speculation that cats moved from competing with humans (7000 BC), through commensality, finally to domestication around 1000 BC (Baldwin, 1975). Studies of mummified cats support the theory that they were domesticated more than 2,500 years ago (Kurushima et al., 2012). Some evidence for cat domestication in the Middle East indicates it may have occurred almost 10,000 years ago (Pickrell, 2004).

The cat genome is not as well studied as that of the dog. Cat breed clusters effectively segregate according to geography, rather than having been bred for behavior, with clades originating from the Mediterranean basin, Europe/America, Asia, and Africa. The Asian cluster was reported to have separated early and expanded in relative isolation, perhaps echoing travel and commerce patterns (Lipinski et al., 2008).

The initial purpose of cats in human lives seems mainly to have centered on vermin control, with no breeding for specific behaviors. Despite humans not selecting stereotypical breed behaviors, several researchers have been using observational and survey tools to understand if behaviors have segregated with the breeds. A survey-based study ranked cat breeds in Japan (Takeuchi & Mori, 2009) for several behaviors. Analysis of the survey data ranked Japanese domestic cats and American shorthairs as the most friendly, playful, and demanding of affection. In the nervousness and timidity behavior group, Chinchilla, Abyssinian, Himalayan, and Siamese were ranked highest.

In addition to survey-based cat breed characterization, open field test methods that measure spontaneous behavior with a focus on emotionality have been used to characterize physiological and behavioral responses in kittens (Marchei et al., 2009). Marchei and colleagues used an open field arena and a novel object to test the response of kittens ages 4 to 10 weeks. The repeated exposures to the novel object in the arena showed that kittens of the Norwegian Forest cat breed were more exploratory than Oriental, Siamese, or Abyssinian breeds. This could be explained by Norwegian Forest kittens being slower to remember prior exposures because of rate of developmental differences between the breeds or genetic differences driving active response in this breed compared with passive response in the Oriental, Siamese, or Abyssinian breeds. Marchei and colleagues (Marchei et al., 2011) expanded their initial analysis by use of a novel and a "threatening" object (a 40-cm cotton case around a metal spring) in the open field test. The response was characterized for the same four breeds, confirming that Norwegian Forest cats avoided the threatening object, suggesting that this breed has an active coping strategy in line with their escape and explore behaviors observed in the previous study. The Oriental, Siamese, and Abyssinian breeds were classified as having a passive response to the threatening object. The study could not rule out a lower arousal threshold in Oriental, Siamese, or Abyssinian breeds.

These cat breed-specific behaviors echo many early studies in dogs that are now being considered in light of DNA polymorphisms in neurology-related genes.

HUMAN BEHAVIORAL GENETICS OFFERS CANDIDATE MECHANISMS OF BEHAVIORS FOR STUDY IN PETS

Behavioral genetics in humans and between species is under intense study, with candidate genes involved in neurology (neurone development, synaptic vesicle formation and efficiency, ion channels, neurotransmitter synthesis and their receptors) being found to have common functional changes.

Behavior-linked candidate gene variation in companion animals is quickly following discoveries made in human and other species. Selected studies have been captured in Table 8.5 to allow the reader to follow authors and behavior type in this quickly moving field.

TABLE 8.5
Companion Animal Genes Under Investigation for Behavioral Links

Gene	Behavior	Ref
Dopamine transporter	Attention deficit/ trainability	Arata et al. (2008); Hejjas, Vas, Kubinyi, et al. (2007); Hejjas, Vas, Topal, et al. (2007)
Dopamine D4 receptor	Impulsivity, attention deficit	Hejjas et al. (2009); Hejjas, Vas, Kubinyi, et al. (2007); Hejjas, Vas, Topal, et al. (2007)
Neuronal/epithelial high affinity glutamate transporter	Activity level	Takeuchi et al. (2009)
Serotonin transporter	No associations to date	Arata et al. (2008)
Tyrosine hydroxylase	Impulsivity	Takeuchi et al. (2005); Kubinyi et al. (2012)

There are several canine compulsive behaviors similar to human conditions that are typified by appearing early in development, seen as repetitive behaviors (tail chasing, flank sucking, and freezing); these usually respond to selective serotonin reuptake inhibitors (SSRI) and have similar imbalance patterns of 5-HT2A and dopamine transporters.

Occasional compulsive tail chasing has been observed in bull terriers and related breeds (Moon-Fanelli & Dodman, 1998), a condition similar to human obsessive-compulsive disorder conditions, which have been linked to serotonin, glutamate, and dopamine neurotransmission. The tail-chasing trait responds to treatment with the tricyclic antidepressant Clomipramine (an SSRI) and is thought to be genetic in bull terriers, Staffordshire bull terriers (Burn, 2011) and German shepherds (Dodman et al., 1996). This trait has been associated with prior boredom, fear, trance, and/or aggression (Dodman et al., 1996; Moon-Fanelli, Dodman, Famula, & Cottam, 2011; Tiira et al., 2012), which could be considered a chain of events like the stalk, chase, and kill sequence described by Coppinger and Coppinger (2001).

Another compulsive behavior, curling up and sucking their flanks, is seen in Doberman pinschers (Moon-Fanelli, Dodman, & Cottam, 2007). Genetic analysis of the trait located a driver of the trait on canine chromosome 7 (Dodman et al., 2010). Dodman and colleagues suggested *Neural Cadherin 2* to be the best candidate gene in this area; it also has been implicated in autism (Wang et al., 2009). *Cadherin 2* has been reported to mediate connections between pre- and postsynaptic membranes, to modulate the efficacy of synaptic transmission, and may play a role in the dynamic switching of adhesive scaffolds in response to synaptic activity (Tanaka et al., 2012). The genetic variant causing flank sucking has not yet been reported.

Although the pet field is benefitting from genes being linked to behaviors in humans, the converse is also now becoming true, as comparative genomic approaches raise the likelihood of scientific discoveries (Lequarré et al., 2011; Ostrander, 2012; Shearin & Ostrander, 2010). Indeed, Ostrander, a leader in canine genetics, suggested that some canine behaviors will be controlled by small numbers of genes allowing quick transfer to human genetic research (Lequarré et al., 2011).

CONCLUSION

The burgeoning area of behavioral genetics in companion animals will continue to make rapid progress because of changes in DNA technology. Both dog and cat genetics are being firmly driven by researchers across the globe with a focus on good sampling techniques and behavioral data collection. The research will aid in our understanding of the behavioral drivers, trainability, and suitability for selection of pets for specific roles such as animal-assisted therapy. The future holds the long-term promise of closer and even more tailored HAI.

REFERENCES

Alvarez, C. E., & Akey, J. M. (2012). Copy number variation in the domestic dog. *Mammalian Genome, 23,* 144–163. http://dx.doi.org/10.1007/s00335-011-9369-8

Arata, S., Ogata, N., Shimozuru, M., Takeuchi, Y., & Mori, Y. (2008). Sequences and polymorphisms of the canine monoamine transporter genes SLC6A2, SLC6A3, and SLC6A4 among five dog breeds. *Journal of Veterinary Medical Science, 70,* 971–975. http://dx.doi.org/10.1292/jvms.70.971

Axelsson, E., Webster, M. T., Ratnakumar, A., Ponting, C. P., Lindblad-Toh, K., & the LUPA Consortium. (2012). Death of *PRDM9* coincides with stabilization of the recombination landscape in the dog genome. *Genome Research, 22,* 51–63. http://dx.doi.org/10.1101/gr.124123.111

Baldwin, J. A. (1975). Notes and speculations in the domestication of the cat in Egypt. *Anthropos, 70,* 428–448.

Bentolila, S., Bach, J.-M., Kessler, J.-L., Bordelais, I., Cruaud, C., Weissenbach, J., & Panthier, J.-J. (1999). Analysis of major repetitive DNA sequences in the dog (*Canis familiaris*) genome. *Mammalian Genome, 10,* 699–705. http://dx.doi.org/10.1007/s003359901074

Berglund, J., Nevalainen, E. M., Molin, A. M., Perloski, M., André, C., Zody, M. C., . . . Webster, M. T., & the LUPA Consortium. (2012). Novel origins of copy number variation in the dog genome. *Genome Biology, 13*(8), R73. http://dx.doi.org/10.1186/gb-2012-13-8-r73

Berry, A., Borgi, M., Terranova, L., Chiarotti, F., Alleva, E., & Cirulli, F. (2012). Developing effective animal-assisted intervention programs involving visiting dogs for institutionalized geriatric patients: A pilot study. *Psychogeriatrics, 12,* 143–150. http://dx.doi.org/10.1111/j.1479-8301.2011.00393.x

Brooks, M. B., Gu, W., Barnas, J. L., Ray, J., & Ray, K. (2003). A Line 1 insertion in the Factor IX gene segregates with mild hemophilia B in dogs. *Mammalian Genome, 14,* 788–795. http://dx.doi.org/10.1007/s00335-003-2290-z

Burn, C. C. (2011). A vicious cycle: A cross-sectional study of canine tail-chasing and human responses to it, using a free video-sharing website. *PLoS ONE, 6*(11), e26553. http://dx.doi.org/10.1371/journal.pone.0026553

Clark, L. A., Tsai, K. L., Starr, A. N., Nowend, K. L., & Murphy, K. E. (2011). A missense mutation in the 20S proteasome β2 subunit of Great Danes having harlequin coat patterning. *Genomics, 97,* 244–248. http://dx.doi.org/10.1016/j.ygeno.2011.01.003

Coppinger, R., & Coppinger, L. (2001). Behavioral conformation. In R. Coppinger & L. Coppinger (Eds.), *Dogs: A startling new understanding of canine origin, behavior and evolution* (pp. 189–224). London, England: Crosskeys Select Books.

Dodman, N. H., Karlsson, E. K., Moon-Fanelli, A., Galdzicka, M., Perloski, M., Shuster, L., . . . Ginns, E. I. (2010). A canine chromosome 7 locus confers compulsive disorder susceptibility. *Molecular Psychiatry, 15,* 8–10. http://dx.doi.org/10.1038/mp.2009.111

Dodman, N. H., Knowles, K. E., Shuster, L., Moon-Fanelli, A. A., Tidwell, A. S., & Keen, C. L. (1996). Behavioral changes associated with suspected complex partial seizures in bull terriers. *Journal of the American Veterinary Medical Association, 208,* 688–091.

Dreger, D. L., & Schmutz, S. M. (2011). A SINE insertion causes the black-and-tan and saddle tan phenotypes in domestic dogs. *The Journal of Heredity, 102*(Suppl. 1), S11–S18. http://dx.doi.org/10.1093/jhered/esr042

Elgier, A. M., Jakovcevic, A., Mustaca, A. E., & Bentosela, M. (2012). Pointing following in dogs: Are simple or complex cognitive mechanisms involved? *Animal Cognition, 15,* 1111–1119. http://dx.doi.org/10.1007/s10071-012-0534-6

Feuerbacher, E. N., & Wynne, C. D. (2012). Relative efficacy of human social interaction and food as reinforcers for domestic dogs and hand-reared wolves. *Journal of the Experimental Analysis of Behavior, 98,* 105–129. http://dx.doi.org/10.1901/jeab.2012.98-105

Fondon, J. W., III, & Garner, H. R. (2004). Molecular origins of rapid and continuous morphological evolution. *Proceedings of the National Academy of Sciences, USA, 101,* 18058–18063. http://dx.doi.org/10.1073/pnas.0408118101

Forman, O. P., Penderis, J., Hartley, C., Hayward, L. J., Ricketts, S. L., & Mellersh, C. S. (2012). Parallel mapping and simultaneous sequencing reveals deletions in BCAN and FAM83H associated with discrete inherited disorders in a domestic dog breed. *PLOS Genetics, 8*(1), e1002462. http://dx.doi.org/10.1371/journal.pgen.1002462

Gácsi, M., McGreevy, P., Kara, E., & Miklósi, Á. (2009). Effects of selection for cooperation and attention in dogs. *Behavioral and Brain Functions*, 5, 31. http://dx.doi.org/10.1186/1744-9081-5-31

Grover, A., & Sharma, P. C. (2014). Development and use of molecular markers: Past and present. *Critical Reviews in Biotechnology*. Advance online publication.

Hejjas, K., Kubinyi, E., Ronai, Z., Szekely, A., Vas, J., Miklósi, Á., . . . Kereszturi, E. (2009). Molecular and behavioral analysis of the intron 2 repeat polymorphism in the canine dopamine D4 receptor gene. *Genes, Brain & Behavior*, 8, 330–336. http://dx.doi.org/10.1111/j.1601-183X.2008.00475.x

Hejjas, K., Vas, J., Kubinyi, E., Sasvari-Szekely, M., Miklósi, Á., & Ronai, Z. (2007). Novel repeat polymorphisms of the dopaminergic neurotransmitter genes among dogs and wolves. *Mammalian Genome*, 18, 871–879. http://dx.doi.org/10.1007/s00335-007-9070-0

Hejjas, K., Vas, J., Topal, J., Szantai, E., Ronai, Z., Szekely, A., . . . Miklósi, Á. (2007). Association of polymorphisms in the dopamine D4 receptor gene and the activity-impulsivity endophenotype in dogs. *Animal Genetics*, 38, 629–633. http://dx.doi.org/10.1111/j.1365-2052.2007.01657.x

Hsu, Y., & Serpell, J. A. (2003). Development and validation of a questionnaire for measuring behavior and temperament traits in pet dogs. *American Veterinary Medical Association Journal*, 223, 1293–1300. http://dx.doi.org/10.2460/javma.2003.223.1293

Hsu, Y., & Sun, L. (2010). Factors associated with aggressive responses in pet dogs. *Applied Animal Behaviour Science*, 123, 108–123. http://dx.doi.org/10.1016/j.applanim.2010.01.013

Imes, D. L., Geary, L. A., Grahn, R. A., & Lyons, L. A. (2006). Albinism in the domestic cat (*Felis catus*) is associated with a tyrosinase (TYR) mutation. *Animal Genetics*, 37, 175–178. http://dx.doi.org/10.1111/j.1365-2052.2005.01409.x

Ishida, Y., David, V. A., Eizirik, E., Schäffer, A. A., Neelam, B. A., Roelke, M. E., . . . Menotti-Raymond, M. (2006). A homozygous single-base deletion in MLPH causes the dilute coat color phenotype in the domestic cat. *Genomics*, 88, 698–705. http://dx.doi.org/10.1016/j.ygeno.2006.06.006

Jakovcevic, A., Elgier, A. M., Mustaca, A. E., & Bentosela, M. (2010). Breed differences in dogs' (*Canis familiaris*) gaze to the human face. *Behavioural Processes*, 84, 602–607. http://dx.doi.org/10.1016/j.beproc.2010.04.003

Jones, P., Chase, K., Martin, A., Davern, P., Ostrander, E. A., & Lark, K. G. (2008). Single-nucleotide-polymorphism-based association mapping of dog stereotypes. *Genetics*, 179, 1033–1044. http://dx.doi.org/10.1534/genetics.108.087866

Kubinyi, E., Vas, J., Hejjas, K., Ronai, Z., Brúder, I., Turcsán, B., . . . Miklósi, Á. (2012). Polymorphism in the tyrosine hydroxylase (TH) gene is associated with activity-impulsivity in German Shepherd Dogs. *PLoS ONE*, 7(1), e30271. http://dx.doi.org/10.1371/journal.pone.0030271

Kurushima, J. D., Ikram, S., Knudsen, J., Bleiberg, E., Grahn, R. A., & Lyons, L. A. (2012). Cats of the pharaohs: Genetic comparison of Egyptian cat mummies to their feline contemporaries. *Journal of Archaeological Science, 39*, 3217–3223. http://dx.doi.org/10.1016/j.jas.2012.05.005

Lequarré, A.-S., Andersson, L., André, C., Fredholm, M., Hitte, C., Leeb, T., . . . Georges, M. (2011). LUPA: A European initiative taking advantage of the canine genome architecture for unravelling complex disorders in both human and dogs. *The Veterinary Journal, 189*, 155–159. http://dx.doi.org/10.1016/j.tvjl.2011.06.013

Li, X., Li, W., Wang, H., Cao, J., Maehashi, K., Huang, L., . . . Brand, J. G. (2005). Pseudogenization of a sweet-receptor gene accounts for cats' indifference toward sugar. *PLOS Genetics, 1*(1), e3. http://dx.doi.org/10.1371/journal.pgen.0010003

Lin, L., Faraco, J., Li, R., Kadotani, H., Rogers, W., Lin, X., . . . Mignot, E. (1999). The sleep disorder canine narcolepsy is caused by a mutation in the hypocretin (orexin) receptor 2 gene. *Cell, 98*, 365–376. http://dx.doi.org/10.1016/S0092-8674(00)81965-0

Lipinski, M. J., Froenicke, L., Baysac, K. C., Billings, N. C., Leutenegger, C. M., Levy, A. M., . . . Lyons, L. A. (2008). The ascent of cat breeds: Genetic evaluations of breeds and worldwide random-bred populations. *Genomics, 91*, 12–21. http://dx.doi.org/10.1016/j.ygeno.2007.10.009

Marchei, P., Diverio, S., Falocci, N., Fatjó, J., Ruiz-de-la-Torre, J. L., & Manteca, X. (2009). Breed differences in behavioural development in kittens. *Physiology & Behavior, 96*, 522–531. http://dx.doi.org/10.1016/j.physbeh.2008.11.015

Marchei, P., Diverio, S., Falocci, N., Fatjó, J., Ruiz-de-la-Torre, J. L., & Manteca, X. (2011). Breed differences in behavioural response to challenging situations in kittens. *Physiology & Behavior, 102*, 276–284. http://dx.doi.org/10.1016/j.physbeh.2010.11.016

Messent, P. R., & Serpell, J. A. (1981). A historical and biological view of the pet owner bond. In B. Fogle (Ed.), *Interactions between people and pets* (pp. 5–22). Springfield, IL: Charles C Thomas.

Minnick, M. F., Stillwell, L. C., Heineman, J. M., & Stiegler, G. L. (1992). A highly repetitive DNA sequence possibly unique to canids. *Gene, 110*, 235–238. http://dx.doi.org/10.1016/0378-1119(92)90654-8

Moon-Fanelli, A. A., & Dodman, N. H. (1998). Description and development of compulsive tail chasing in terriers and response to clomipramine treatment. *Journal of the American Veterinary Medical Association, 212*, 1252–1257.

Moon-Fanelli, A. A., Dodman, N. H., & Cottam, N. (2007). Blanket and flank sucking in Doberman pinschers. *Journal of the American Veterinary Medical Association, 231*, 907–912. http://dx.doi.org/10.2460/javma.231.6.907

Moon-Fanelli, A. A., Dodman, N. H., Famula, T. R., & Cottam, N. (2011). Characteristics of compulsive tail chasing and associated risk factors in bull terriers. *Journal of the American Veterinary Medical Association, 238*, 883–889. http://dx.doi.org/10.2460/javma.238.7.883

Nagasawa, M., Tsujimura, A., Tateishi, K., Mogi, K., Ohta, M., Serpell, J. A., & Kikusui, T. (2011). Assessment of the factorial structures of the C-BARQ in Japan. *Journal of Medical Veterinary Science, 73,* 869–875. http://dx.doi.org/10.1292/jvms.10-0208

Ostrander, E. A. (2012). Both ends of the leash—the human links to good dogs with bad genes. *The New England Journal of Medicine, 367,* 636–646. http://dx.doi.org/10.1056/NEJMra1204453

Parker, H. G., VonHoldt, B. M., Quignon, P., Margulies, E. H., Shao, S., Mosher, D. S., . . . Ostrander, E. A. (2009). An expressed fgf4 retrogene is associated with breed-defining chondrodysplasia in domestic dogs. *Science, 325*(5943), 995–998. http://dx.doi.org/10.1126/science.1173275

Pecon-Slattery, J., Pearks Wilkerson, A. J., Murphy, W. J., & O'Brien, S. J. (2004). Phylogenetic assessment of introns and SINEs within the Y chromosome using the cat family *felidae* as a species tree. *Molecular Biology and Evolution, 21,* 2299–2309. http://dx.doi.org/10.1093/molbev/msh241

Pickrell, J. (2004, April 8). Oldest known pet cat? 9,500-year-old burial found on Cyprus. *National Geographic News.* Retrieved from http://news.nationalgeographic.com/news/2004/04/0408_040408_oldestpetcat.html

Schröder, C., Bleidron, C., Hartmann, S., & Tiedemann, R. (2009). Occurrence of Can-SINEs and intron sequence evolution supports robust phylogeny of pinniped carnivores and their terrestrial relatives. *Gene, 448*(2), 221–226.

Shearin, A. L., & Ostrander, E. A. (2010). Leading the way: Canine models of genomics and disease. *Disease Models & Mechanisms, 3,* 27–34. http://dx.doi.org/10.1242/dmm.004358

Takeuchi, Y., Hashizume, C., Arata, S., Inoue-Murayama, M., Maki, T., Hart, B. L., & Mori, Y. (2009). An approach to canine behavioural genetics employing guide dogs for the blind. *Animal Genetics, 40,* 217–224. http://dx.doi.org/10.1111/j.1365-2052.2008.01823.x

Takeuchi, Y., Hashizume, C., Chon, E. M., Momozawa, Y., Masuda, K., Kikusui, T., & Mori, Y. (2005). Canine tyrosine hydroxylase (TH) gene and dopamine beta-hydroxylase (DBH) gene: Their sequences, genetic polymorphisms, and diversities among five different dog breeds. *The Journal of Veterinary Medical Science, 67,* 861–867. http://dx.doi.org/10.1292/jvms.67.861

Takeuchi, Y., & Mori, Y. (2009). Behavioral profiles of feline breeds in Japan. *The Journal of Veterinary Medical Science, 71,* 1053–1057. http://dx.doi.org/10.1292/jvms.71.1053

Tanaka, H., Takafuji, K., Taguchi, A., Wiriyasermkul, P., Ohgaki, R., Nagamori, S., . . . Kanai, Y. (2012). Linkage of N-cadherin to multiple cytoskeletal elements revealed by a proteomic approach in hippocampal neurons. *Neurochemistry International, 61,* 240–250. http://dx.doi.org/10.1016/j.neuint.2012.05.008

Tiira, K., Hakosalo, O., Kareinen, L., Thomas, A., Hielm-Björkman, A., Escriou, C., . . . Lohi, H. (2012). Environmental effects on compulsive tail chasing in dogs. *PLoS ONE, 7*(7), e41684. http://dx.doi.org/10.1371/journal.pone.0041684

Udell, M. A., Dorey, N. R., & Wynne, C. D. (2010). What did domestication do to dogs? A new account of dogs' sensitivity to human actions. *Biological Reviews of the Cambridge Philosophical Society, 85,* 327–345. http://dx.doi.org/10.1111/j.1469-185X.2009.00104.x

van den Berg, L., Schilder, M. B. H., de Vries, H., Leegwater, P. A. J., & van Oost, B. A. (2006). Phenotyping of aggressive behavior in golden retriever dogs with a questionnaire. *Behavior Genetics, 36,* 882–902. http://dx.doi.org/10.1007/s10519-006-9089-0

Wang, K., Zhang, H., Ma, D., Bucan, M., Glessner, J. T., Abrahams, B. S., . . . Hakonarson, H. (2009). Common genetic variants on 5p14.1 associate with autism spectrum disorders. *Nature, 459,* 528–533. http://dx.doi.org/10.1038/nature07999

Webster, J. (Ed.). (2011). *Management and welfare of farm animals: The UFAW farm handbook* (5th ed.). London, England: Wiley-Blackwell.

Zeng, R., Farias, F. H., Johnson, G. S., McKay, S. D., Schnabel, R. D., Decker, J. E., . . . O'Brien, D. P. (2011). A truncated retrotransposon disrupts the GRM1 coding sequence in Coton de Tulear dogs with Bandera's neonatal ataxia. *Journal of Veterinary Internal Medicine, 25,* 267–272. http://dx.doi.org/10.1111/j.1939-1676.2010.0666.x

9

ADVANCING THE SOCIAL NEUROSCIENCE OF HUMAN–ANIMAL INTERACTION: THE ROLE OF SALIVARY BIOSCIENCE

NANCY A. DRESCHEL AND DOUGLAS A. GRANGER

Although the field of human–animal interaction (HAI) has been recognized for decades, there have been recent calls to increase the number of sound scientific studies, including the integration of valid and repeatable biobehavioral measures of both short- and long-term effects (e.g., Esposito, McCune, Griffin, & Maholmes, 2011). New technologies and techniques have increased the opportunities for developmental, health, and therapeutic investigations in varied human populations, as well as the study of the welfare implications of HAI on animals. As investigators move to include these techniques in their research, it is important that they understand the potential applications and limitations of each measure. In recent decades, the opportunity to explore biosocial relationships in humans and animals has been facilitated

In the spirit of full disclosure, DAG is the chief scientific and strategy advisor of Salimetrics, LLC (State College, PA) and SalivaBio LLC (Baltimore, MD) and these relationships are managed by the policies of the Conflict of Interest Committee at the Johns Hopkins University School of Medicine and the Office of Research Integrity and Adherence at Arizona State University.

http://dx.doi.org/10.1037/14856-012
The Social Neuroscience of Human–Animal Interaction, L. S. Freund, S. McCune, L. Esposito, N. R. Gee, and P. McCardle (Editors)

by the ability to measure inter- and intraindividual differences in the activity of biological systems such as the hypothalamic–pituitary–adrenal (HPA) axis, hypothalamic–pituitary–gonadal (HPG) axis, and autonomic nervous system (ANS) noninvasively in oral fluids (saliva). In fact, technical innovations in the field of salivary bioscience reveal that information may be obtained from oral fluid specimens about the activity of a broad array of physiological systems, pathogen and chemical exposures, and genetic variability relevant to basic biological function, health, and disease. The attention saliva has received as a research biospecimen is largely due to the perceptions of sample collection as quick, uncomplicated, cost-efficient, minimally invasive, and acceptable to children and parents, as well as to animals and their handlers.

The purpose of this chapter is to describe best practices and provide a road map to enable investigators interested in the social neuroscience of HAI to integrate the tools of salivary bioscience into their conceptual and measurement models. Although we focus on a variety of analytes in oral fluid, we refer readers specifically interested in vasopressin and oxytocin to Chapter 5 by Carter and Porges and Chapter 6 by Beetz and Bales in this volume.

THEORETICAL AND CONCEPTUAL ISSUES

The study of hormones and behavior is the scientific foundation for most research using salivary analytes in the social, behavioral, and health sciences. Modern thought assumes that biological changes influence behavior and vice versa, enabling individuals to respond flexibly and fluidly to changes in the environment. Gottlieb (1992) elaborated and suggested that certain physiological processes are activated only when components of the "behavioral surface" are unable to accommodate the challenge. In his model, the first line of "adaptation" and the most flexible and fluid mechanisms available to an individual to adjust to changes in the environment involve coping resources, change in behavior, restructuring cognitions and perceptions of the event, or all of the above. Thus, physiological systems sensitive to context (i.e., psychobiology of stress) are activated only when the adjustment cannot be handled by the behavioral surface. In theory, the activation of these physiological subsystems adjusts over repeated encounters with the same situation or circumstances. Habituation is essential to maintain homeostasis and the integrity of the organism, whereas repeated or chronic activation of the HPA or ANS has deleterious consequences via effects on multiple body systems (McEwen, 1998). The relation of activity and regulation of the biological systems to behavior is assumed to be dependent on the social context. To paraphrase Sapolsky (2005), hormones do not cause behavior; they increase the probability that existing behavioral tendencies will be expressed given the right circumstances. Related to this, it has been suggested that developmental mechanisms calibrate activation thresholds and response magnitudes within environmentally responsive biological systems to match ecological conditions encountered in life, rendering certain individuals intrinsically more or less biologically sensitive and susceptible to context (e.g., Ellis, Boyce, Belsky, Bakermans-Kranenburg, & van IJzendoorn, 2011).

These theoretical issues suggest that examining the associations and dissociations between concurrent activity of biological systems in relation to behavior, health, and social relationships is critical. Also, advancing our understanding of how social and contextual forces influence the coordination of these systems may provide insight into how individual differences in physiological reactivity and regulation contribute to social, behavioral, and cognitive processes and well-being. Viewed from this set of assumptions and perspectives, HAIs represent a specialized context in which biobehavioral relationships may be modified or changed. Since Friedmann's landmark study of pet ownership and cardiac patient survival (Friedmann, Katcher, Lynch, & Thomas, 1980), much HAI research has focused on physiological measures (e.g., stress hormone levels, sympathetic nervous system activation, and cardiovascular reactivity) to examine the effects of animals on their human companions. Although HAI refers to the interaction and effects of each species on the other, fewer investigators have examined their bidirectional effects. Salivary analytes have the potential to advance measurement models that aim to examine these complex reciprocal and context-dependent influences both in humans and in animals.

HUMAN SALIVARY BIOSCIENCE

To integrate biology, behavior, and context into theoretical and analytical models, biospecimens must be collected (a) repeatedly from the same individuals, (b) without causing burden or stress to the human or animal, and (c) in a variety of settings. Historically, the initial wave of studies using salivary analytes (those prior to the late 1990s) often ignored key facts about oral biology and the nature of saliva as a biospecimen. This may have compromised the value of the information gained. Biobehavioral research on HAI can benefit from that unfortunate history by reviewing and understanding some of the fundamentals of oral biology related to the special characteristics of oral fluid.

Oral Fluid

The "saliva" specimen is actually a composite of oral fluids secreted from many different glands (Veerman, van den Keybus, Vissink, & Nieuw

Amerongen, 1996). The major source glands are located in the upper posterior area of the oral cavity (*parotid gland* area), lower area of the mouth between the cheek and jaw (*submandibular gland* area), and under the tongue (*sublingual gland* area). A small fraction of oral fluid (crevicular fluid) also comes from serum leakage in the cleft area between each tooth and its surrounding gums or via leakage from serum due to mucosal injury or inflammation.

Each secretory gland produces a fluid that differs in volume, composition, and constituents (Veerman et al., 1996), thus each source gland's contribution to the pool of oral fluid varies. For instance, *mucins* make saliva viscous, elastic, and sticky to protect tooth enamel against wear and to encapsulate microorganisms. These glycoproteins are not present in oral fluid secreted by the parotid gland. Under resting conditions—when there is minimal fluid contribution from the parotid gland and the levels of mucins in saliva are high and consequential—specimens will be more viscous (Nieuw Amerongen, Bolscher, & Veerman, 1995).

Oral fluid is water-like in composition and has a pH (acidity) range between 6 and 9. Foods and substances placed in the mouth are capable of changing salivary acidity very quickly because the fluid has minimal buffering capacity. Immunoassays are a method of choice for assaying many salivary analytes. The antibody–antigen binding step during an immunoassay is compromised when the specimen is highly acidic (pH < 3) or basic (pH > 9). This unique characteristic of saliva interacts with procedures used to collect it and can compromise measurement accuracy.

Many of the salivary analytes used in biobehavioral studies (e.g., steroid hormones) are serum constituents transported into saliva either by *filtration* through the tight spaces between acini (duct cells in the salivary glands) or *diffusion* through acinar membranes (Vining, McGinley, & Symons, 1983). Some of the analytes found in oral fluids (e.g., enzymes, mucins, cystatins, histatins) are synthesized, stored, and released from the granules within the secretory cells of the saliva glands. Still others are components of humoral immunity (antibodies, complement) or signaling molecules (cytokines, chemokines, growth factors) secreted by cells of the mucosal immune system. Furthermore, saliva collected using procedures common in biobehavioral studies contains sufficient cellular material to obtain a high quantity and quality DNA. An understanding of whether an analyte is transported into oral fluid by filtration or passive diffusion, secreted from salivary glands, or released or derived from cells locally in the oral mucosa is essential to interpreting the meaning of individual differences in that measure.

The secretion of oral fluids is influenced by several factors: the day–night cycle, chewing movement of the mandibles, taste and smell, iatrogenic effects of medications that cause xerostoma (dry mouth) and medical interventions (radiation), and conditions (e.g., Sjögren's syndrome) that affect salivary gland function (Atkinson et al., 1990). Salivary glands are directly inner-vated by ANS nerves (e.g., Garrett, 1987), and activation of the sympa-thetic and parasympathetic components of the ANS response to stress decrease or increase saliva flow rates, respectively. The levels of salivary analytes produced in the mouth, like alpha-amylase (sAA) and secretory IgA (SIgA), and the levels of those that migrate into saliva from blood by fil-tration through the junctions between duct cells in the salivary gland (e.g., dehydroepiandrosterone-sulfate [DHEA-S] and other conjugated steroids) are influenced by the rate of saliva secretion (e.g., Kugler, Hess, & Haake, 1992). For these saliva analytes, a correction must be made by multiplying the measured concentration or activity of the analyte (e.g., U/ml, pg/ml, µg/dl) by the flow rate (ml/min) to express the measure as *output as a function of time* (e.g., U/min, pg/min, µg/min; e.g., .50 µg/dl × 0.5 ml/min = 1 µg/min), or at a minimum, flow rate (ml/min) should be used as a covariate in the statistical analyses.

The U.S. Centers for Disease Prevention and Control (the CDC) notes that unless visibly contaminated with blood, oral fluid is not a Class II bio-hazard. This statement has contributed to the perception that saliva is *safer* to work with than blood. In reality, other than obviously not needing needles during the collection steps, this may be something of a misperception. Even under normative-healthy conditions, more than 250 species of bacteria are present in oral fluids (Paster et al., 2001). During upper respiratory infections, oral fluids are highly likely to contain agents of disease. Oral fluid specimens should be handled like all Class II biohazards with *universal precautions* in both research and diagnostic applications.

Sample Collection

In the past, saliva collection devices have involved cotton-based absor-bent materials. Cotton placed in the mouth for 2 to 3 minutes is rapidly saturated by oral fluids, which are then expressed into collection vials by centrifugation or compression. Most of the time, this is convenient, simple, and time efficient. However, when the absorbent capacity is large and sam-ple volume small, the specimen absorbed can be diffusely distributed in the cotton fibers, making sample recovery problematic, with possibly higher rates of missing data and artificially low cortisol estimates. This absorption pro-cess has the potential to interfere with the immunoassay of several salivary analytes.

Early studies used serum assays modified for use with saliva by, among other things, requiring large saliva test volumes (200–400 µl). To collect suffi-cient test volumes, saliva flow was often stimulated via chewing (gums, dental wax) or tasting (sugar crystals, powdered drink mixes, citric acid drops) substances. When not used minimally and/or consistently, some of these methods are capable of changing immunoassay performance (e.g., Schwartz, Granger, Susman, Gunnar, & Laird, 1998). Indirectly, stimulants also influence measurement of the levels of salivary analytes dependent on saliva flow rate (SIgA, DHEA-S, neuropeptide Y, vasoactive intestinal peptide). We advise avoidance of these techniques unless pilot studies show that their application does not adversely affect measurement validity of the salivary analytes of interest.

Given this quick review of the sources of oral fluids, it is not surprising that studies show the placement of oral swabs in the mouth has the potential to introduce variation in the measured levels or activity of some salivary analytes. Depending on where in the mouth an absorbent device is placed, a different fluid type may be collected, and if not controlled, may contribute to measurement error across sampling occasions within and between subjects. Caution must be exercised to minimize this threat to measurement validity by standardizing instructions and monitoring compliance.

Collecting *whole saliva* by passive drool can minimize these threats to validity (Granger et al., 2007). Briefly, participants are asked to imagine that they are chewing their favorite food, slowly move their jaws in a chewing motion, and allow the oral fluid to pool in their mouth without swallowing. Next, they gently force the specimen through a short device (e.g., SalivaBio LLC, Carlsbad, California) into a vial. There are several advantages of this procedure: (a) a large sample volume (0.5–1.5 ml) can be collected within a short collection time frame (3–5 minutes), (b) target collection volume can be confirmed by visual inspection in the field, (c) the fluid collected is a pooled specimen mixture of the output from all salivary glands, (d) it does not introduce interference related to stimulating or absorbing saliva, and (e) samples can be aliquoted and archived for future assays.

Most techniques that have been studied have unique benefits as well as shortcomings that prevent universal application. When possible, saliva collection methods should always be piloted in the field to ensure that they do not contribute to measurement error, in relation to the exact assay protocols to be used.

Blood Leakage Into Oral Fluid

To meaningfully index *systemic* biological activity, quantitative estimates of an analyte (e.g., hormone) in saliva may need to be highly correlated with the levels measured in serum. The magnitude of this serum–saliva association depends, in part, on consistency in the processes used to transport circulating molecules into oral fluids. When the integrity of diffusion or filtration is compromised, the level of the serological marker in saliva will be affected. Most serum constituents are present in serum in much higher levels (10–100 fold) than in saliva.

Blood and blood products can leak into oral fluids via burns, abrasions, or cuts to the cheek, tongue, or gums. Blood in oral fluid is more prevalent among individuals who suffer from poor oral health (i.e., open sores, periodontal disease, gingivitis), endure certain infectious diseases, or engage in behavior known to influence oral health negatively.

Spiking whole blood into saliva reveals that samples visibly contaminated with blood will present varying degrees of yellow-brownish hue. A simple 5-point Blood Contamination in Saliva Scale (BCSS; Kivlighan et al., 2004) offers the following response options: (a) "Saliva appears clear, no visible color"; (b) "Saliva has a hint of color, a little brown or yellow tint is barely visible"; (c) "Saliva has a clearly visible yellow or brown tint"; (d) "Yellow or brown coloring is more than just a tint, color is obvious but not very deep"; and (e) "Saliva is very colored, deep, rich, dark yellow or brown is very apparent" (pp. 41–42). Under healthy conditions, BCSS ratings ($N = 42$) averaged 1.33; after microinjury caused by vigorous tooth brushing, ratings averaged 2.42.

In the context of research on HAI, (a) participants should be screened for events in their recent history that could cause blood leakage into saliva by asking questions related to oral health (i.e., "Do your gums bleed when you floss or brush your teeth?"), shedding teeth, or open sores or injury to the oral cavity; (b) sampling saliva within 45 minutes of microinjury to the oral cavity (e.g., brushing teeth) should be avoided (Kivlighan et al., 2004); and (c) samples should be systematically inspected at the collection point and, if visibly contaminated with blood, excluded from analyses.

Particulate Matter and Interfering Substances

The integrity of oral fluid samples can also be influenced by items placed in the mouth. Food residue in the oral cavity after drinking or eating may change salivary pH or composition (viscosity) and/or contain substances (e.g., bovine hormones, active ingredients in medications, enzymes) that cross-react in immune- or kinetic-reaction assays. We recommend a simple solution: Research participants should not consume food or drink within the 20 minutes prior to sample donation. If anything has been eaten within this time window, participants should rinse their mouth with water prior to providing a specimen. It is important to note, however, that they must wait at least 10 minutes after drinking before a specimen is collected to avoid diluting it with water and artificially lowering concentration/volume (µg/dl, ng/ml, pg/ml) or activity/volume (U/ml) estimates of salivary analytes.

Access to food and drink should be carefully planned and scheduled when study designs involve repeated sample collections over long time periods.

Sample Handling, Transport, and Storage

Typically, once specimens are collected, they should be kept cold or frozen to prevent degradation of some salivary analytes and to restrict the activity of proteolytic enzymes and growth of bacteria. How samples are handled, stored, and transported has the potential to influence sample integrity and measurement validity. Our recommendation is conservative: After collection, saliva samples should be kept frozen (at least −20 °C), or at a minimum kept cold (on ice or refrigerated) until they can be frozen that day, and repeated freeze–thaw cycles should be avoided. Also note that some salivary analytes (e.g., neuropeptides) may require specimens to be treated with inhibitors (e.g., ethylenediaminetetraacetic acid [EDTA], aprotinin) or flash frozen on the spot to minimize rapid degradation.

Medications

We recommend that the name, dosage, and schedule of all prescription and over-the-counter medications taken within the past 48 hours be recorded in the field. This information should be used (covaried, controlled) to statistically rule out the possibility that medication use is driving the primary salivary analyte-outcome relationships of interest.

ANIMAL SALIVARY BIOSCIENCE

Oral fluid has been collected and analyzed for a variety of compounds in many species. Although the principles of salivary bioscience in other animals are similar to those in humans, there are species-specific considerations that should be taken into account when planning and carrying out saliva collection in animals. As oral fluid varies in production, composition, and function, investigators should familiarize themselves with the species with which they are dealing.

In herbivores, such as the horse, saliva plays a particularly important role in combining with food to create a bolus and provide lubrication for swallowing. Like humans, horses have parotid, mandibular, and sublingual salivary glands. Smaller glands in the labial, buccal, and lingual regions also provide moisture for the area in which they are found (Schummer, Nickel, & Sack, 1979). Horses produce large quantities of saliva (10–12 L/day), which, like that of humans and canines, is a hypotonic solution. Unlike dogs and humans, the parotid reflex of horses is not conditioned to external stimuli, that is, salivary secretion does not begin until food is actually prehended and the animal begins to eat (Alexander & Hickson, 1969).

In addition to the parotid, mandibular, and sublingual salivary glands, dogs and cats have a zygomatic salivary gland located in the upper jaw behind the eye. The pH of dog saliva varies between 7.34 and 7.8, whereas that of cat saliva averages around 7.5 (National Research Council, 2006). Canine saliva also has some buffering capabilities to protect individuals against ingestion of more acidic substances. Dogs and cats (as well as marine mammals, dolphins, and sea lions) lack the salivary alpha-amylase enzyme that is found in very low concentration in the equine. Unlike the situation for many other species, an important function of canine saliva is evaporative cooling; the flow rate of saliva increases greatly in response to increased environmental and body temperature (Swenson & Reece, 1993).

Sample Collection

A number of investigators have successfully collected and analyzed oral fluids in companion animal species involved in animal-assisted therapy and activities, including horses and dogs (Dreschel & Granger, 2005, 2009; Peeters, Sulon, Beckers, Ledoux, & Vandenheede, 2011). Although it is possible to collect saliva from domestic cats (Siegford, Walshaw, Brunner, & Zanella, 2003), it is difficult to obtain adequate sample volume to process. Collection devices must be used to collect oral fluid from animals, as no technique for collecting passive drool in common companion animals has been developed. A primary consideration in oral fluid collection from animals is safety to the researcher; because saliva collection requires close proximity of the handler's fingers to the animal's mouth, it should not be attempted on aggressive, fearful, or anxious dogs. Many of the same limitations (e.g., poor recovery of fluid from cotton fibers, interference of cotton with salivary analytes) reported in humans also apply to animals, so it is recommended that an inert material, such as methylcellulose, which has not been shown to interfere with immunoassay of cortisol, be used for sample collection (Dreschel & Granger, 2009). Some researchers have used salivary stimulants such as citric acid, but the danger of interference with salivary analytes exists in animals as in humans (Dreschel & Granger, 2009; Schwartz et al., 1998); as with human sampling, we recommend avoiding these unless pilot studies show that they do not adversely affect the measurement validity of the compounds of interest. Likewise, flavoring of the collection device has been shown to interfere with salivary cortisol analysis (Dreschel & Granger, 2009).

The technique we have found to be most useful in collecting saliva from dogs is as follows: (a) one handler holds the dog steady to avoid it walking away, while the other firmly holds a 5-in. saliva collection swab in one hand and lifts the subject's lip with the other hand; (b) place the swab into the mouth through the space between the upper and lower canine teeth and the premolars, which often causes the dog to open his or her mouth; (c) hold the muzzle loosely shut so that the dog can chew on the swab, stimulating saliva flow; and (d) hold the swab firmly and move it around to sample from several areas of the mouth. It may take 2 to 3 minutes for saturation of the swab. The swab can also be used to wipe saliva from the areas between the gingiva and cheeks of the dog. Training using positive reinforcement can facilitate oral fluid collection procedures. After saturation, the swab can be compressed through a 5- to 10-cc syringe to extract the saliva or placed in a storage tube. Handling postcollection is similar to that of human samples. This technique has been successful for veterinarians, veterinary technicians, researchers, and dog owners (Dreschel, 2007; Dreschel & Granger, 2005, 2009).

Blood Leakage, Particulate Matter, and Interfering Substances

As with humans, the contamination of oral fluid with blood and other substances can be a limitation with animal sample collection. In domestic canines, the presence of dental disease and gingivitis will increase the risk of blood contamination and should be avoided or at least noted. Additionally, dogs often retain food between their teeth or eat or chew on other substances (grass, dirt, sticks, rawhide bones) that could interfere with salivary analytes, so dogs should not eat for 30 minutes before sample collection. Although it is recommended that researchers reward the dog immediately after sample collection, if a follow-up sample will be collected, the reward should be a small treat that will not require much chewing. Likewise, because dogs retain fluid in their mouth after drinking, there is a risk of sample dilution if the sample is taken too soon after the dog drinks.

Medications and Neutering Status

Many dogs take monthly heartworm and flea preventatives. Corticosteroids and other anti-inflammatory drugs are also commonly prescribed. Although the influence of these on salivary analytes has not been well documented, any medications taken should be recorded. In addition, it is important to record age, gender, and neutering status of animals used in HAI research; the role of spaying and castration on salivary hormone levels has not been researched.

RESEARCH DESIGNS, SAMPLING SCHEMES, AND ANALYTICAL STRATEGY

In this section, we describe the logic behind some common research designs and saliva sampling schemes and describe several analytical strategies for these designs.

Basal Levels

A *basal level* is the level of activity of an analyte that represents the "stable state" of the host during a resting period. One approach to assessing basal levels has been to sample early in the morning, before the events of the day are able to contribute variation. Levels of salivary analytes may be influenced by inherent moment-to-moment, diurnal, and/or monthly variation in their production/release, rate of their metabolism/degradation, and sensitivity to environmental influences and whether they are measured quantitatively or qualitatively. Given these issues, a single time-point measure of salivary analytes (other than invariant genetic polymorphisms), except under very unique circumstances, is unlikely to yield meaningful results for basal levels. The minimally invasive nature of oral fluid collection enhances the reliability of basal estimates of salivary analytes by sampling at the same time of day across a number of sampling days, then aggregating across days. Theoretically, the more inherent variation in the analyte, the more days of sampling would be required.

Stress Reactivity and Regulation to Acute Events

The vast majority of studies have used research designs that test time-dependent changes in salivary analytes (i.e., cortisol, sAA) following exposure to a discrete event. The number of samples collected depends on the specific analyte, questions being addressed, tolerance for sampling burden by participants, and logistical and practical issues. The optimal design for the measurement of salivary cortisol and sAA reactivity and regulation involves a pre-pre-[task]-post-post-post-postsampling scheme with samples collected on arrival to the lab (after consent); immediately before the task (after a period of relaxation); then again immediately, 5, 20, and 40 minutes postchallenge.

Although some developmental studies have yielded consistent mean-level differences in salivary analytes before and after exposure to a stressful or novel event, there are generally wide-ranging intraindividual differences in stress-related reactivity. Some individuals will exhibit unexpected patterns of change, including no change, as well as continuous increases or decreases in analyte levels at least during the time period in which the analyte was sampled.

Person-Oriented Approach for Identifying Reactivity and Recovery

Studies using pre- and posttask saliva sampling designs have also explored individual differences in reactivity or recovery. Early studies often classified youth as cortisol reactors or nonreactors on the basis of a 10% to 15% difference between pre- and posttask levels (e.g., Susman, Dorn, Inoff-Germain, Nottelmann, & Chrousos, 1997). The logic was that a difference of this size was 2 to 3 times larger than the intra-assay coefficient of variation (CV; 3%–5%). The intra-assay CV reflects the error inherent in the assay by comparing results from the same samples assayed twice, with an intra-assay CV equal to 0 meaning perfect reliability. For cortisol, we added the criterion of an absolute difference of at least .02 µg/dl, as this value is 2 to 3 times higher than the lower limit of our salivary cortisol assay's sensitivity (i.e., the smallest value distinguishable from zero is 0.007 µg/dl). The next step typically involved either multivariate logistic regression or discriminant function analyses to predict reactor status (e.g., Granger, Weisz, McCracken, Ikeda, & Douglas, 1996). The limitation of this simple approach is that all reactors are grouped together, even though some may only show a minor increase and others may show substantial change. This approach becomes complex when multiple poststress samples are collected and the focus is on individual differences in the trajectory of reactivity and recovery over time. For these designs, latent growth modeling approaches, such as growth mixture modeling, may be used to identify homogenous subpopulations within a larger heterogeneous population and for the identification of meaningful groups of individuals with specific growth trajectories. Given the wide-ranging individual differences in physiological responses to stress, continued efforts to use this advanced type of individual-oriented approach seem valuable.

Patterns of Reactivity and Recovery Across Multiple Occasions

Consistent patterns of HPA or ANS activation across time or situations (e.g., high or low reactivity across conditions) may be especially informative when investigating individual differences in risk or resilience. Before much progress can be made on this front, a consensus is needed regarding the best manner by which to group individuals into these different patterns of reactivity and recovery. Growth mixture modeling could be a useful tactic to identify these patterns and profiles.

Diurnal Rhythm

An important component of variability within individuals in salivary analyte levels is the diurnal rhythm of production (e.g., Gunnar & Vazquez, 2001). In humans, most salivary hormone levels (e.g., cortisol) are high in the morning, decline before noon, and then decline more slowly in the afternoon and evening hours (Nelson, 2005). By contrast, levels of sAA show the reversed pattern with low levels in the morning and higher levels in the afternoon (Nater, Rohleder, Schlotz, Ehlert, & Kirschbaum, 2007). The nonlinear nature of these patterns requires multiple sampling time points to create adequate statistical models. A typical sampling design for salivary cortisol and sAA involves sampling immediately on waking, 30 minutes postwaking, midday (around noon), in the late afternoon, and immediately prior to bed (Hellhammer et al., 2007; Nater et al., 2007).

Horses have also been shown to have a diurnal secretion of cortisol, with levels higher in the morning (0600 hours) and lower in the evening (1800 hours; van der Kolk, Nachreiner, Schott, Refsal, & Zanella, 2001). Interestingly, although episodic secretion of cortisol has been shown in both cats and dogs, a diurnal rhythm to cortisol secretion has not been identified (Kemppainen & Peterson, 1996; Koyama, Omata, & Saito, 2003).

Measurement of Momentary Biobehavioral Associations in Everyday Contexts

Documenting everyday events and emotions that help explain changes in analyte levels or activity across the time period of interest may strengthen causal inference when these assessments are paired with samples across multiple days. Recent advances in information technology (computerized handheld devices, such as PDAs) have made these self-assessments of momentary emotions and events possible during the course of individuals' everyday lives (e.g., Stone et al., 2003). Research designs typically involve diary-sample pairings several times per day and across multiple days. In studies focusing on cortisol, saliva samples are collected approximately 20 minutes after each diary entry.

Although much HAI research has concentrated on therapeutic interventions, more recent work has focused on the general effects of living with and working with animals on daily human lives (e.g., Allen, Blascovich, & Mendes, 2002). The role of animal influence on specific populations continues to be examined (e.g., Aydin et al., 2012; Viau et al., 2010). Using noninvasive physiological measures such as salivary analytes in any population can add to the richness of these data. In addition, the role of human influence on the welfare of animals in shelters (e.g., Bergamasco et al., 2010); in competition (Jones & Josephs, 2006); in therapy work (Glenk et al., 2014); and in military, police, and search-and-rescue working situations (Haverbeke, Diederich, Depiereux, & Giffroy, 2008; Horváth, Dóka, & Miklósi, 2008) has been examined using canine salivary cortisol. Because of the relative ease of canine saliva collection, handlers, pet owners, and researchers are able to collect samples at home and during training and working sessions. This research will help to identify and ameliorate stress or other negative outcomes that could be associated with the inclusion of animals in therapeutic and working situations.

Associative Relations of Salivary Analytes Between Dyads

In several studies, salivary analyte levels (e.g., cortisol, testosterone, sAA) are associated in dating couples (Powers, Pietromonaco, Gunlicks, & Sayer, 2006), newlywed couples (Cohan, Booth, & Granger, 2003), siblings (Schreiber et al., 2006), and parent–child dyads (Sethre-Hofstad, Stansbury, & Rice, 2002). Constructs related to these patterns of symmetry have varied substantially across studies. Studies have found dyadic physiological symmetry to be associated with negative correlates such as exposure to domestic violence and harsh parenting practices (Hibel, Granger, Blair, & Cox, 2009), maternal depression (Laurent, Ablow, & Measelle, 2011), marital dissatisfaction (Saxbe & Repetti, 2010), and shared negative affect (Papp, Pendry, & Adam, 2009). Yet other studies relate dyadic physiological symmetry to positive correlates such as friendship strength (Goldstein, Field, & Healy, 1989), and maternal sensitivity (van Bakel, & Riksen-Walraven, 2008), making it difficult to form decisive conclusions about the implications of this coordination. One area of discrepancy in the examination of physiological symmetry is the type of methods used in analyses.

The associative dyadic relationship between animals and humans is an underdeveloped area of research in HAI. Most research to date has focused on the effects of animals on humans (particularly in animal-assisted therapy and activities), or the influence of humans on animals (in the context of animal welfare), but few studies have looked at the association of each species in biobehavioral terms and the dyadic relationships of this bidirectional influence. Odendaal and Meintjes (2003) presented some of the first research examining this dyadic relationship using neurophysiological correlates of positive human–dog interactions on the basis of blood sampling and arterial blood pressure measurement. Handlin et al. (2011) also examined the physiological response to interaction between dogs and their owners using blood sampling. The use of salivary analytes in such studies allows for research beyond the laboratory into other areas of HAI. Jones and Josephs (2006) examined the change in salivary cortisol in agility dogs, relative to their handlers' basal levels and changes in testosterone in winning and losing teams. Lit, Boehm, Marzke, Schweitzer, and Oberbauer (2010) compared human salivary cortisol and testosterone with dogs' pulse and body temperature during search-and-rescue certification tests.

ANALYTES IN SALIVA OF INTEREST TO RESEARCH ON HAI

To date, the range of salivary analytes that have been integrated into studies of HAI has been restricted relative to the possibilities. Many may not know that the National Institute for Dental and Craniofacial Research (NIDCR) initiated a multisite program project charged with characterizing the salivary proteome in humans. The list includes more than 1,000 analytes (Hu, Loo, & Wong, 2007). Salivary analytes vary in terms of how they can be interpreted, which influences their *value* to HAI research.

Some analytes are present in saliva because oral fluid represents an ultrafiltrate of serum constituents. This group of analytes has high value because their levels in saliva are highly correlated with and reflect levels in general circulation. These measures enable investigators to make inferences about systemic physiological states. Adrenal and gonadal hormones are exemplars of this category of salivary markers (see Exhibit 9.1). Cortisol or corticosterone (in relevant species) is the most common analyte measured in animal saliva as a physiological marker of stress.

The majority of analytes in oral fluid are produced locally in the oral cavity and secreted from salivary glands; their levels may reflect features of and variations in oral biology rather than systemic physiology. Many salivary immune and inflammatory markers such as neopterin, beta-2-microglobulin, and cytokines (see Exhibit 9.1) fall into this category. SIgA in domestic canines is a potentially useful marker of stress (Kikkawa, Uchida, Nakade, & Taguchi, 2003). Individual differences may represent systemic immune function or status, but a more likely major contributor is local inflammatory processes related to oral health and disease.

A subset of analytes is produced locally by salivary glands, but the levels vary predictably with systemic physiological activation. The activation of ANS affects the release of catecholamines from nerve endings, and these compounds' action on adrenergic receptors influences the activity of the salivary glands. sAA is considered a *surrogate marker* of ANS activation, with the majority of findings linking it to sympathetic activation via beta-adrenergic pathways in humans. Cats and dogs do not produce salivary amylase, so this is of no value in these species (Dreschel, 2007). Salivary measures of neuropeptide Y and vasoactive intestinal peptide may also serve as surrogate markers of ANS.

EXHIBIT 9.1
Salivary Analytes of Potential Interest to Human–Animal Interaction

Endocrine	*Nucleic acid*
Aldosterone	Human genomic
Androstenedione	Mitochondrial
Cortisol	Bacterial
Dihydroepiandrosterone, and –sulfate	mRNA
Estradiol, estriol, esterone	Microbial
Progesterone, 17-OH progesterone	Viral
Testosterone	
Melatonin	*Antibodies specific for antigens*
	Measles
Immune/inflammation	Mumps
Secretory immunoglobulin A (SIgA)	Rubella
Neopterin	Hepatitis A,B,C,E
Soluble tumor necrosis factor receptors	Herpes simplex
Beta-2-microglobulin (B_2M)	Epstein-Barr
Cytokines, chemokines	HIV
C-reactive protein (CRP)	CMV
Autonomic nervous system	
Alpha-amylase (sAA)	
Vasoactive intestinal peptide (VIP)	
Neuropeptide Y (NPY)	

Antibodies to specific antigens are also measurable in oral fluids. Antibodies to human immunodeficiency virus and hepatitis C are the exemplars in this category of salivary analytes, and Exhibit 9.1 offers several additional examples. The presence of an antibody in oral fluids reflects immunological history of pathogen/microbe exposure and, depending on the specific antibody measured, may represent local and/or systemic immune activity or current or prior exposure. A variety of pharmaceuticals, abused substances, and environmental contaminants can be quantitatively monitored in oral fluids. Cotinine, a metabolite of nicotine, is routinely measured in oral fluid to estimate primary and secondary exposure to nicotine. Urinary cotinine has been previously measured in dogs and cats to measure the effects of environmental tobacco smoke (Bertone-Johnson, Procter-Gray, Gollenberg, Ryan, & Barber, 2008; McNiel et al., 2007).

Within the recent past, technical advances confirm that high quantity and quality DNA can be extracted from whole saliva. Genetic polymorphisms can be determined from the same specimens already in use, or planned for use, to assess individual differences in salivary analytes and biomarkers. Our preliminary studies suggest that global and specific methylation assays are technically feasible using DNA extracted from cells in oral fluid, raising the possibility that saliva-based measurements may contribute to study epigenetic phenomena in research related to HAI.

CONCLUSION

Sampling oral fluid is minimally invasive, collection is simple and discrete, and specimens can be collected repeatedly without interrupting the flow of social interaction. As the number of substances that can be reliably measured increases, oral fluid may become a biospecimen of choice for studies of HAI. These technical advances enable the construction and evaluation of measurement models related to how individual differences in several integrated biological systems are related to behavior, cognition, psychopathology, and health and how these differences moderate the effect of HAI on subsequent psychosocial and behavioral adjustment in humans and animals. With careful attention to the special issues noted here, salivary bioscience has the potential to profoundly impact our understanding of the social neuroscience of HAI.

REFERENCES

Alexander, F., & Hickson, J. C. D. (1969). The salivary and pancreatic secretions of the horse. In A. T. Phillipson (Ed.), *Physiology of digestion and metabolism in the ruminant* (pp. 375–389). Newcastle Upon Tyne, England: Oriel Press.

Allen, K., Blascovich, J., & Mendes, W. B. (2002). Cardiovascular reactivity and the presence of pets, friends, and spouses: The truth about cats and dogs. *Psychosomatic Medicine, 64*, 727–739.

Atkinson, J. C., Travis, W. D., Pillemer, S. R., Bermudez, D., Wolff, A., & Fox, P. C. (1990). Major salivary gland function in primary Sjögren's syndrome and its relationship to clinical features. *The Journal of Rheumatology, 17*, 318–322.

Aydin, N., Krueger, J. I., Fischer, J., Hahn, D., Kastenmuller, A., Frey, D., & Fischer, P. (2012). "Man's best friend": How the presence of a dog reduces mental distress after social exclusion. *Journal of Experimental Social Psychology, 48*, 446–449. http://dx.doi.org/10.1016/j.jesp.2011.09.011

Bergamasco, L., Osella, M. C., Savarino, P., Larosa, G., Ozella, L., Manassero, M., . . . Re, G. (2010). Heart rate variability and saliva cortisol assessment in shelter dog: Human-animal interaction effects. *Applied Animal Behaviour Science, 125*, 56–68. http://dx.doi.org/10.1016/j.applanim.2010.03.002

Bertone-Johnson, E. R., Procter-Gray, E., Gollenberg, A. L., Ryan, M. B., & Barber, L. G. (2008). Environmental tobacco smoke and canine urinary cotinine level. *Environmental Research, 106*, 361–364. http://dx.doi.org/10.1016/j.envres.2007.09.007

Cohan, C. L., Booth, A., & Granger, D. A. (2003). Gender moderates the relationship between testosterone and marital interaction. *Journal of Family Psychology, 17*, 29–40. http://dx.doi.org/10.1037/0893-3200.17.1.29

Dreschel, N. A. (2007). *The biobehavioral effects of stress related to fear and anxiety in domestic canines*. Unpublished doctoral dissertation, Pennsylvania State University.

Dreschel, N. A., & Granger, D. A. (2005). Physiological and behavioral reactivity to stress in thunderstorm-phobic dogs and their caregivers. *Applied Animal Behaviour Science, 95*, 153–168. http://dx.doi.org/10.1016/j.applanim.2005.04.009

Dreschel, N. A., & Granger, D. A. (2009). Methods of collection for salivary cortisol measurement in dogs. *Hormones and Behavior, 55*, 163–168. http://dx.doi.org/10.1016/j.yhbeh.2008.09.010

Ellis, B. J., Boyce, W. T., Belsky, J., Bakermans-Kranenburg, M. J., & van IJzendoorn, M. H. (2011). Differential susceptibility to the environment: An evolutionary–neurodevelopmental theory. *Development and Psychopathology, 23*, 7–28. http://dx.doi.org/10.1017/S0954579410000611

Esposito, L., McCune, S., Griffin, J. A., & Maholmes, V. (2011). Directions in human-animal interaction research: Child development, health, and therapeutic interventions. *Child Development Perspectives, 5*, 205–211. http://dx.doi.org/10.1111/j.1750-8606.2011.00175.x

Friedmann, E., Katcher, A. H., Lynch, J. J., & Thomas, S. A. (1980). Animal companions and one-year survival of patients after discharge from a coronary care unit. *Public Health Reports, 95*, 307–312.

Garrett, J. R. (1987). The proper role of nerves in salivary secretion: A review. *Journal of Dental Research, 66*, 387–397. http://dx.doi.org/10.1177/00220345870660020201

Glenk, L. M., Kothgassner, O. D., Stetina, B. U., Palme, R., Kepplinger, B., & Baran, H. (2014). Salivary cortisol and behavior in therapy dogs during animal-assisted interventions: A pilot study. *Journal of Veterinary Behavior, 9*, 98–106. http://dx.doi.org/10.1016/j.jveb.2014.02.005

Goldstein, S., Field, T., & Healy, B. (1989). Concordance of play behavior and physiology in preschool friends. *Journal of Applied Developmental Psychology, 10*, 337–351. http://dx.doi.org/10.1016/0193-3973(89)90034-8

Gottlieb, G. (1992). *Individual development and evolution: The genesis of novel behavior*. Oxford, England: Oxford University Press.

Granger, D. A., Kivlighan, K. T., Fortunato, C., Harmon, A. G., Hibel, L. C., Schwartz, E. B., & Whembolua, G. L. (2007). Integration of salivary biomarkers into developmental and behaviorally-oriented research: Problems and solutions for collecting specimens. *Physiology & Behavior, 92*, 583–590. http://dx.doi.org/10.1016/j.physbeh.2007.05.004

Granger, D. A., Weisz, J. R., McCracken, J. T., Ikeda, S. C., & Douglas, P. (1996). Reciprocal influences among adrenocortical activation, psychosocial processes, and the behavioral adjustment of clinic-referred children. *Child Development, 67*, 3250–3262. http://dx.doi.org/10.2307/1131777

Gunnar, M. R., & Vazquez, D. M. (2001). Low cortisol and a flattening of expected daytime rhythm: Potential indices of risk in human development. *Development and Psychopathology, 13,* 515–538. http://dx.doi.org/10.1017/S0954579401003066

Handlin, L., Hydbring-Sandberg, E., Nilsson, A., Ejdeback, M., Jansson, A., & Uvnas-Moberg, K. (2011). Short-term interaction between dogs and their owners: Effects on oxytocin, cortisol, insulin and heart rate—an exploratory study. *Anthrozoös, 24,* 301–315. http://dx.doi.org/10.2752/175303711X13045914865385

Haverbeke, A., Diederich, C., Depiereux, E., & Giffroy, J. M. (2008). Cortisol and behavioral responses of working dogs to environmental challenges. *Physiology & Behavior, 93,* 59–67. http://dx.doi.org/10.1016/j.physbeh.2007.07.014

Hellhammer, J., Fries, E., Schweisthal, O. W., Schlotz, W., Stone, A. A., & Hagemann, D. (2007). Several daily measurements are necessary to reliably assess the cortisol rise after awakening: State- and trait components. *Psychoneuroendocrinology, 32,* 80–86. http://dx.doi.org/10.1016/j.psyneuen.2006.10.005

Hibel, L. C., Granger, D. A., Blair, C., & Cox, M. J. (2009). Intimate partner violence moderates the association between mother-infant adrenocortical activity across an emotional challenge. *Journal of Family Psychology, 23,* 615–625. http://dx.doi.org/10.1037/a0016323

Horváth, Z., Dóka, A., & Miklósi, Á. (2008). Affiliative and disciplinary behavior of human handlers during play with their dog affects cortisol concentrations in opposite directions. *Hormones and Behavior, 54,* 107–114. http://dx.doi.org/10.1016/j.yhbeh.2008.02.002

Hu, S., Loo, J. A., & Wong, D. T. (2007). Human saliva proteome analysis. *Annals of the New York Academy of Sciences, 1098,* 323–329. http://dx.doi.org/10.1196/annals.1384.015

Jones, A. C., & Josephs, R. A. (2006). Interspecies hormonal interactions between man and the domestic dog (*Canis familiaris*). *Hormones and Behavior, 50,* 393–400. http://dx.doi.org/10.1016/j.yhbeh.2006.04.007

Kempppainen, R. J., & Peterson, M. E. (1996). Domestic cats show episodic variation in plasma concentrations of adrenocorticotropin, alpha-melanocyte-stimulating hormone (alpha-MSH), cortisol and thyroxine with circadian variation in plasma alpha-MSH concentrations. *European Journal of Endocrinology, 134,* 602–609. http://dx.doi.org/10.1530/eje.0.1340602

Kikkawa, A., Uchida, Y., Nakade, T., & Taguchi, K. (2003). Salivary secretory IgA concentrations in beagle dogs. *The Journal of Veterinary Medical Science, 65,* 689–693. http://dx.doi.org/10.1292/jvms.65.689

Kivlighan, K. T., Granger, D. A., Schwartz, E. B., Nelson, V., Curran, M., & Shirtcliff, E. A. (2004). Quantifying blood leakage into the oral mucosa and its effects on the measurement of cortisol, dehydroepiandrosterone, and testosterone in saliva. *Hormones and Behavior, 46,* 39–46. http://dx.doi.org/10.1016/j.yhbeh.2004.01.006

Koyama, T., Omata, Y., & Saito, A. (2003). Changes in salivary cortisol concentrations during a 24-hour period in dogs. *Hormone and Metabolic Research, 35*, 355–357. http://dx.doi.org/10.1055/s-2003-41356

Kugler, J., Hess, M., & Haake, D. (1992). Secretion of salivary immunoglobulin A in relation to age, saliva flow, mood states, secretion of albumin, cortisol, and catecholamines in saliva. *Journal of Clinical Immunology, 12*, 45–49. http://dx.doi.org/10.1007/BF00918272

Laurent, H. K., Ablow, J. C., & Measelle, J. (2011). Risky shifts: How the timing and course of mothers' depressive symptoms across the perinatal period shape their own and infant's stress response profiles. *Development and Psychopathology, 23*, 521–538. http://dx.doi.org/10.1017/S0954579411000083

Lit, L., Boehm, D., Marzke, S., Schweitzer, J., & Oberbauer, A. M. (2010). Certification testing as an acute naturalistic stressor for disaster dog handlers. *Stress: The International Journal of the Biology of Stress, 13*, 392–401. http://dx.doi.org/10.3109/10253891003667896

McEwen, B. S. (1998). Protective and damaging effects of stress mediators. *The New England Journal of Medicine, 338*, 171–179. http://dx.doi.org/10.1056/NEJM199801153380307

McNiel, E. A., Carmella, S. G., Heath, L. A., Bliss, R. L., Le, K.-A., & Hecht, S. S. (2007). Urinary biomarkers to assess exposure of cats to environmental tobacco smoke. *American Journal of Veterinary Research, 68*, 349–353. http://dx.doi.org/10.2460/ajvr.68.4.349

Nater, U. M., Rohleder, N., Schlotz, W., Ehlert, U., & Kirschbaum, C. (2007). Determinants of the diurnal course of salivary alpha-amylase. *Psychoneuroendocrinology, 32*, 392–401. http://dx.doi.org/10.1016/j.psyneuen.2007.02.007

National Research Council. (2006). *Nutrient requirements of dogs and cats.* Washington, DC: National Academies Press.

Nelson, R. J. (2005). *An introduction to behavioral endocrinology.* Sunderland, MA: Sinauer Associates.

Nieuw Amerongen, A. V., Bolscher, J. G. M., & Veerman, E. C. I. (1995). Salivary mucins: Protective functions in relation to their diversity. *Glycobiology, 5*, 733–740. http://dx.doi.org/10.1093/glycob/5.8.733

Odendaal, J. S. J., & Meintjes, R. A. (2003). Neurophysiological correlates of affiliative behaviour between humans and dogs. *Veterinary Journal, 165*, 296–301. http://dx.doi.org/10.1016/S1090-0233(02)00237-X

Papp, L. M., Pendry, P., & Adam, E. K. (2009). Mother-adolescent physiological synchrony in naturalistic settings: Within-family cortisol associations and moderators. *Journal of Family Psychology, 23*, 882–894. http://dx.doi.org/10.1037/a0017147

Paster, B. J., Boches, S. K., Galvin, J. L., Ericson, R. E., Lau, C. N., Levanos, V. A., . . . Dewhirst, F. E. (2001). Bacterial diversity in human subgingival plaque. *Journal of Bacteriology, 183*, 3770–3783. http://dx.doi.org/10.1128/JB.183.12.3770-3783.2001

Peeters, M., Sulon, J., Beckers, J.-F., Ledoux, D., & Vandenheede, M. (2011). Comparison between blood serum and salivary cortisol concentrations in horses using an adrenocorticotropic hormone challenge. *Equine Veterinary Journal, 43*, 487–493. http://dx.doi.org/10.1111/j.2042-3306.2010.00294.x

Powers, S. I., Pietromonaco, P. R., Gunlicks, M., & Sayer, A. (2006). Dating couples' attachment styles and patterns of cortisol reactivity and recovery in response to a relationship conflict. *Journal of Personality and Social Psychology, 90*, 613–628.

Sapolsky, R. M. (2005). The influence of social hierarchy on primate health. *Science, 308*, 648–652. http://dx.doi.org/10.1126/science.1106477

Saxbe, D., & Repetti, R. L. (2010). For better or worse? Coregulation of couples' cortisol levels and mood states. *Journal of Personality and Social Psychology, 98*, 92–103. http://dx.doi.org/10.1037/a0016959

Schreiber, J. E., Shirtcliff, E., Van Hulle, C., Lemery-Chalfant, K., Klein, M. H., Kalin, N. H., . . . Goldsmith, H. H. (2006). Environmental influences on family similarity in afternoon cortisol levels: Twin and parent-offspring designs. *Psychoneuroendocrinology, 31*, 1131–1137. http://dx.doi.org/10.1016/j.psyneuen.2006.07.005

Schummer, A., Nickel, R., & Sack, W. O. (1979). *The viscera of the domestic animals*. New York, NY: Springer-Verlag.

Schwartz, E. B., Granger, D. A., Susman, E. J., Gunnar, M. R., & Laird, B. (1998). Assessing salivary cortisol in studies of child development. *Child Development, 69*, 1503–1513. http://dx.doi.org/10.1111/j.1467-8624.1998.tb06173.x

Sethre-Hofstad, L., Stansbury, K., & Rice, M. A. (2002). Attunement of maternal and child adrenocortical response to child challenge. *Psychoneuroendocrinology, 27*, 731–747. http://dx.doi.org/10.1016/S0306-4530(01)00077-4

Siegford, J. M., Walshaw, S. O., Brunner, P., & Zanella, A. J. (2003). Validation of a temperament test for domestic cats. *Anthrozoös, 16*, 332–351. http://dx.doi.org/10.2752/089279303786991982

Stone, A. A., Broderick, J. E., Schwartz, J. E., Shiffman, S., Litcher-Kelly, L., & Calvanese, P. (2003). Intensive momentary reporting of pain with an electronic diary: Reactivity, compliance, and patient satisfaction. *Pain, 104*, 343–351. http://dx.doi.org/10.1016/S0304-3959(03)00040-X

Susman, E. J., Dorn, L. D., Inoff-Germain, G., Nottelmann, E. D., & Chrousos, G. P. (1997). Cortisol reactivity, distress behavior, and psychological problems in young adolescents: A longitudinal perspective. *Journal of Research on Adolescence, 7*, 81–105. http://dx.doi.org/10.1207/s15327795jra0701_5

Swenson, M. J., & Reece, W. O. (1993). *Dukes' physiology of domestic animals* (11th ed., pp. 349–351). Ithaca, NY: Cornell University Press.

van Bakel, H. J. A., & Riksen-Walraven, J. M. (2008). Adrenocortical and behavioral attunement in parents with 1-year-old infants. *Developmental Psychobiology, 50*, 196–201. http://dx.doi.org/10.1002/dev.20281

van der Kolk, J. H., Nachreiner, R. F., Schott, H. C., Refsal, K. R., & Zanella, A. J. (2001). Salivary and plasma concentration of cortisol in normal horses and horses with Cushing's disease. *Equine Veterinary Journal, 33,* 211–213. http://dx.doi.org/10.1111/j.2042-3306.2001.tb00604.x

Veerman, E. C. I., van den Keybus, P. A., Vissink, A., & Nieuw Amerongen, A. V. (1996). Human glandular salivas: Their separate collection and analysis. *European Journal of Oral Sciences, 104,* 346–352. http://dx.doi.org/10.1111/j.1600-0722.1996.tb00090.x

Viau, R., Arsenault-Lapierre, G., Fecteau, S., Champagne, N., Walker, C.-D., & Lupien, S. (2010). Effect of service dogs on salivary cortisol secretion in autistic children. *Psychoneuroendocrinology, 35,* 1187–1193. http://dx.doi.org/10.1016/j.psyneuen.2010.02.004

Vining, R. F., McGinley, R. A., & Symons, R. G. (1983). Hormones in saliva: Mode of entry and consequent implications for clinical interpretation. *Clinical Chemistry, 29,* 1752–1756.

10

FROM THE DOG'S PERSPECTIVE: WELFARE IMPLICATIONS OF HAI RESEARCH AND PRACTICE

NANCY R. GEE, KARYL J. HURLEY, AND JOHN M. RAWLINGS

Human–animal interaction (HAI) refers to the mutual and dynamic relationships between people and animals and the ways in which these interactions may affect physical and psychological health and well-being (McCardle, McCune, Griffin, & Esposito, 2011). The effects described from HAI investigations are often focused largely on the many well-known positive benefits for people: for example, reduction of blood pressure, lower risk of allergies and asthma in children, reduced stress and depression. Historically, few HAI investigations have intentionally focused on or addressed the potential benefits or harm of HAI to the animals involved.

The intent of this chapter is to provide a balanced perspective on animal well-being and welfare in the study of HAI and, in the context of this volume, to explore the value of neuroscience as an avenue for discovering the basis of the human–animal bond. Neuroscience may be able to provide explanations for some of the observed effects of HAI, as well as identify the

http://dx.doi.org/10.1037/14856-013
The Social Neuroscience of Human–Animal Interaction, L. S. Freund, S. McCune, L. Esposito, N. R. Gee, and P. McCardle (Editors)

underlying neurobiological mechanisms for people's attachment to animals. As investigators search for suitable HAI biomarkers, such as oxytocin, cortisol, and functional magnetic resonance imaging (fMRI) patterns of activation, whether in the setting of a research facility or in practical animal-assisted interventions, it is imperative both to the animal and to the validity of the research to ensure that the welfare and well-being of the animals are adequately addressed.

In this chapter, we define animal welfare and the implications of involving animals in HAI research, focusing our discussion specifically on dogs. Although many animal species have been involved in HAI, dogs are the most commonly studied species. Perhaps this is due to their popularity as pets and the relative ease in training them to display predictable behaviors. Dogs are the oldest domesticated species and have evolved to be particularly sensitive to human cues.

We have a moral and legal obligation to ensure the welfare of animals in our care. Although specific animal cruelty/welfare laws and policies vary from region to region, we posit that a higher standard should be set and upheld by the HAI community of scholars and practitioners. Poor welfare can lead to illness and suffering, which is unacceptable and detracts from any potential gain from HAI. Previous studies have shown positive correlations between measures of animal welfare (e.g., heart-rate variability) and capability in guide dogs and search and rescue dogs. Healthy, less-stressed dogs learn faster and produce better results (for a review of these studies, see Rooney, Gaines, & Hiby, 2009). Finally, animal welfare is of increasing importance as pets are seen more as family members, and with closer scrutiny by animal advocates, it benefits all those working with animals to ensure their welfare standards are high.

WHAT IS ANIMAL WELFARE?

Defining *animal welfare* continues to exercise the minds of philosophers who consider the moral obligations of humankind toward animals; of practitioners who work with the animals; and of authorities who have to find an acceptable playing field for all interested parties, including the animals themselves. "The state of the animal as it attempts to cope with its environment" is a concept put forward by Broom (1986, 1988) to define the welfare of an animal and is perhaps highly relevant in the context of this book. In the study of HAI we are placing animals in a variety of situations in which they may not be in control of their environment, and because they are not necessarily the subject of study, they may not be afforded the safeguards provided to "experimental animals" by regulatory authorities in traditional research settings. However, the status of an animal, its needs and its quality of life, should not, we might argue, be dictated by whether it is the subject of an experiment or a participant within a program of study.

What are the needs of animals? The concept of Five Freedoms originated with the Report of the Technical Committee to Enquire into the Welfare of Animals Kept Under Intensive Livestock Husbandry Systems (Brambell and The Technical Committee, 1965). The authors of this report set out to define ideal states rather than standards of welfare; after further refinement by the Farm Animal Welfare Council, these ideal states are described as follows: (a) freedom from hunger and thirst; (b) freedom from discomfort; (c) freedom from pain, injury, or disease; (d) freedom to express normal behavior; and (e) freedom from fear and distress.

Sandoe and Christiansen (2008) went beyond describing animal welfare in terms of basic states by attempting to define what might be considered a good animal life. They pointed out that it is insufficient to consider merely how humans feel and advocate taking into account the fact that animals have needs and preferences that are different from those of humans. Again, this may have relevance to the field of HAI research, in which the subject of the study is often the person, and thus human needs and preferences are considered as primary. More recently the idea of "a life worth living," particularly in relation to farm animal welfare, has been proposed (Farm Animal Welfare Council, 2009). This concept builds on the holistic idea of an animal's welfare throughout its whole life, based on more fundamental concepts such as basic states, overall welfare, and quality and value of life (Yeates, 2011). In providing a framework in which the concept could be considered, Yeates (2011) concluded that fewer animals might have lives worth avoiding, should a life worth living be considered. The concept is as relevant to companion animals, both in the domestic and research environments, as it is to farm animal welfare. However, the practicalities of assessing an animal's experiences during individual HAI encounters as well as throughout its lifetime is not easy, although later in this chapter we will see how one can begin to identify the telltale signs of the animal's mental state and some strategies for helping the individual animal cope with its environment.

ANIMALS IN RESEARCH

Although a broad range of animals have been used in HAI research, the predominant species is the dog. This probably reflects our special relationship with dogs, the practicality of using dogs within different settings, and their trainability. However, dogs also command special representation within regulated research environments, especially within the recently published

European Union [EU] Directive on the Protection of Animals Used for Research (European Parliament and Council of the European Union, 2010). Together with cats and horses, dogs are given special status that is only marginally below that of nonhuman primates. No rationale for this segregation from other species is provided, but the result of the United Kingdom (UK) consultation with stakeholders in the transposition of the EU Directive into UK law (United Kingdom Home Office, 2012) has overwhelmingly endorsed it. The 1966 Animal Welfare Act in the United States (U.S. Department of Agriculture, 2013) gives the dog equal status to other warm-blooded mammals, but the Act excludes other common research animals such as birds, mice, and rats.

Typically, studies of HAI will involve consideration by Institutional Review Boards or their equivalents to ensure the welfare of the human participants. However, where an animal is not the subject of a study, often there is no obligation to involve the Institutional Animal Care and Use Committee (IACUC) or the equivalent Animal Welfare Body (AWB) where these are stipulated by regulations. Of course, where the animal does become the subject of the study, it is imperative that an IACUC/AWB be consulted irrespective of whether the animal is housed in the institution that initiates the research or is privately owned. We would argue, however, that the IACUC/AWB should always be consulted, even when the animal is not the subject of the research study. Serpell, Coppinger, and Fine (2006) pointed out a number of examples in which dogs will enter unfamiliar surroundings and encounter a wide range of people, yet they are rarely asked if they are motivated to interact socially with unfamiliar humans given a choice! Perhaps then, the IACUC/AWB should ask the investigator whether the selection of the dog for the proposed study has taken into account the animal's predisposition to interacting with unfamiliar people.

The housing, socialization, and habituation of animals, and dogs in particular, have a significant impact on the success of the study. Indeed, when Russell and Burch (1959) put forward the concept of the "3Rs" (reduction, refinement, replacement), they commented that the most humane treatment possible for experimental animals, "far from being an obstacle, is actually a prerequisite for successful animal experiments" (p. 3). In Europe at least, regulatory requirements for housing of dogs in research environments have moved a long way toward meeting the needs of the dogs through adoption of the EU Directive (European Parliament and Council of the European Union, 2010) by harmonizing, on a pan-European level, the housing and care standards.

However, the subjects of socialization and habituation remain less well defined in any regulatory context. The work of Hubrecht, Serpell, and Poole (1992) sets some of the first expectations of what environmental enrichment can look like in a kennel environment. Others have built on this by defining new standards for dog housing and husbandry (Loveridge, 1998) and describing both animate and inanimate enrichment opportunities for dogs (Moesta, McCune, Deacon, & Kruger, 2015; Wells, 2004). Research studies continue to be published on the role of different enrichment tools in research and shelter environments (Pullen, Merrill, & Bradshaw, 2010, 2012a, 2012b), demonstrating our desire to better understand the needs of dogs.

Habituation to experimental procedures is as important as socialization with people and other animals and, if done well, can promote more of a willing partnership between the researcher and the animal. For example, if we are trying to measure the cortisol response in a dog in response to stroking, we will be unsuccessful if the act of taking the sample is a stressor in its own right. Habituation to mouth handling through positive reinforcement techniques as the dog grows from puppy to adult will allow collection of saliva samples by having the dog willingly chew on a piece of cotton wool, from which the saliva can be extracted (Rawlings, Deacon, & Healey, 2007; see also Chapter 9, this volume). This is more likely to provide a true reflection of the animal's state in response to the treatment, for example, the stroking effect. Typically habituation takes time to develop but can result in more reliable data: Training dogs to remain motionless can eliminate extraneous effects such as air movement when taking measurements of water loss from the surface of the skin (Watson, Fray, Clarke, Yates, & Markwell, 2002). Berns, Brooks, and Spivak (2012) used positive reinforcement techniques rather than general anesthesia to encourage two pet dogs to remain still within an fMRI scanner, allowing the study of conscious brain activity; for their success, they credited the reliance on the willing partnership between researcher and dog.

ANIMALS IN PRACTICE

Most frequently, HAI research takes place with animals in applied settings involving animal-related activities or targeted therapeutic interventions. It is essential in all of these encounters that the selected animals remain calm and unstressed and display behaviors that indicate that they are comfortable in their surroundings and with the people with whom they are interacting. In this section, we focus on how to protect the welfare of the individual dog by choosing a temperamentally suited animal, ensuring proper training, minimizing stressful encounters, and maintaining the health of the dog.

Defining the Settings

Animal-assisted intervention (AAI), according to Kruger and Serpell (2010), refers to "any intervention that intentionally includes or incorporates animals as part of a therapeutic or ameliorative process or milieu" (p. 36). The use of the term *AAI* is therefore flexible enough to include programs that fit within a therapeutic approach (animal-assisted therapy) and those that aim only to positively impact the lives of the participants (animal-assisted activities). For ease of discussion, we will use the broader term AAI going forward.

Animal-assisted activities (AAAs), according to the Pet Partners (formally Delta Society) website, provide "opportunities for motivational, educational, recreational, and/or therapeutic benefits to enhance quality of life. AAAs are delivered in a variety of environments by specially trained professionals, para-professionals, and/or volunteers, in association with animals that meet specific criteria" (http://www.petpartners.org/AAA-Tinformation). Key features include a lack of specific treatment goals, no requirement that treatment providers and volunteers take detailed notes, and spontaneous visit content.

Animal-assisted therapy (AAT), according to the same source, is "a goal-directed intervention in which an animal that meets specific criteria is an integral part of the treatment process. AAT is directed and/or delivered by a health/human service professional with specialized expertise, and within the scope of practice of his/her profession" (http://www.petpartners.org/AAA-Tinformation). Key features include clearly specified goals and objectives for each individual, and progress is typically tracked and measured.

It is important to note that service animals (e.g., guide dogs for the blind) represent a category distinct from animals used in AAI. They are not included in the previous definitions primarily because the Americans With Disabilities Act of 1990 (ADA; 1991) states that the role of the service animal is to perform some function or task that individuals cannot perform themselves due to their disability (U.S. Department of Justice, 2010). The ADA provides disabled individuals and their service dogs with legal access to a wide variety of public buildings, businesses, hospitals, and restaurants. It is important to note that this legal protection does not extend to animals involved in AAI, AAA, or AAT.

Behavioral Expectations of the Dog

There are a number of organizations (see Table 10.1 for three examples) that now certify or register therapy dogs (TDs). This is a generic term used to describe dogs working in AAI, although not all of these dogs work in therapeutic settings. The process of certification varies from organization to organization, but in general the dog and handler team must pass a test that is typically an extension of the American Kennel Club Canine Good Citizen (CGC) test. For the CGC test, dogs must be able to walk on a loose leash through a crowd of people; sit, lie down, and stay on command; react appropriately to other dogs and distractions; come when called; accept a friendly stranger;

TABLE 10.1
Relevant Resources

Topic	Source
Getting started in AAI	
Video guide to AAT with a focus on the dog	*Lending a helping paw: A guide to animal-assisted therapy*, McConnell (2012); http://www.patriciamcconnell.com
Overview of AAT with a focus on ethical issues	*Therapy dogs today*, Butler (2004) (See also information in this table on TD certifying agencies and training)
Identifying stress in dogs	
Video examples of stress	*Canine behavior: Observing and interpreting canine body postures*, Suzanne Hetts & Daniel Epstein; http://animalbehaviorassociates.com/program-canine-behavior-posture.htm
	The language of dogs, Kalnajs (2007)
Understanding dog behavior	*Canine body language: A photographic guide*, Aloff (2005)
	How to speak dog, Coren (2000)
	The culture clash, Donaldson (1996)
	For the love of the dog, McConnell (2006); http://www.patriciamcconnell.com
	On talking terms with dogs: Calming signals, Turid (2005)
Temperament test information	
Review of	Jones & Gosling (2005)
Validity of	Barnard, Siracusa, Reisner, Valsecchi, & Serpell (2012)
TD certifying agencies	
Pet Partners/Delta Society	http://www.petpartners.org
Therapy Dogs International	http://www.tdi-dog.org
Therapy Dogs Inc.	http://www.therapydogs.com
Training resources	
Positive/clicker training	Karen Pryor; http://www.clickertraining.com
General training	Ian Dunbar; http://www.siriuspup.com
	Suzanne Clothier; http://www.suzanneclothier.com
TD specific training	*Therapy dogs: Training your dog to reach others*, Davis (2002)

Note. TD = therapy dog, AAI = animal-assisted intervention, AAT = animal-assisted therapy.

and sit politely for petting. TDs are also typically required to react calmly and positively to children, medical equipment, patient infirmities (unusual gait, stilted movements, or the use of cane, crutches, walker, or wheelchair), and loud noises and have an understanding of commands such as "leave it" (ignore interesting items such as food) and "go visit" (go greet a stranger in a polite manner). TDs are typically trained to accept a wide variety of types of touch from the people with whom they will interact. In general, the dogs must be predictable and controllable under distracting circumstances, keep all four paws on the floor unless specifically asked for another behavior, be calm in chaos, be nonreactive to other dogs or loud noises, and be comfortable during a period of supervised separation from their handler.

Once certified, TDs' access to public buildings, hospitals, schools, and other facilities is at the discretion of the management of that institution or business. The certification process itself does not provide any additional legal benefits to the dog/handler team, but it is usually accompanied by an insurance policy, provided by the certifying agency, designed to cover potential liability claims related to volunteer AAIs. It is important for handlers to work closely with the facility management or staff to have a clear understanding of the unique rules and policies.

Steps to AAI Success for Dogs and Humans

Selecting the Dog for AAI Work

Success in developing a nonstressed, calm therapy dog must begin with the selection of an appropriate candidate. When selecting a dog for TD certification and AAI work, people often focus on characteristics such as breed, size, and type of coat (long vs. short hair, or shedding vs. nonshedding) or where to obtain the dog (breeder vs. shelter). In fact, the success of an AAI dog is neither breed- nor size-dependent; the most important characteristic is temperament. Other characteristics of the dog are also relevant, such as overall health and longevity, and it is likely that canine cognition will also be an important consideration in selecting dogs for this work.

Admittedly, certain AAI settings can be better suited for smaller dogs, so given specific circumstances, size can be relevant. Similarly, because nonshedding breeds tend to release far fewer hairs and thus release less dander (the actual allergen), they may be less likely to stimulate an allergic reaction; however, they do still release dander and are thus not truly "hypo-allergenic." In all dogs, good grooming practices, including bathing and brushing prior to AAIs, will vastly reduce shedding and dander as potential allergy problems, as will a well-ventilated environment for AAI work.

Temperament is the true predictor of a dog's suitability for AAI. In her 2012 video, *Lending a Helping Paw: A Guide to Animal Assisted Therapy* (see Table 10.1), McConnell details a number of characteristics that make a good TD. The dog should be affiliative with strangers and friendly. She points out that many such dogs will make extensive direct eye contact with humans, and people seem to enjoy this. Because not all breeds of dog are comfortable with direct eye contact but can still make excellent TDs, that particular quality cannot universally be considered a suitability criterion. The dog should be physically calm, psychologically sound and nonreactive, comfortable in a variety of settings, emotionally mature, and healthy. Contraindications include being overly enthusiastic, not affiliative (aloof to people), fearful, resource guarding (e.g., protective of food or toys), owner focused, licking or drooling, or aggressive in any context. McConnell points out that although some dogs can be definitely ruled out as candidates for TD work, other young, enthusiastic dogs might simply need to mature into the role, becoming good TD candidates later in their lifetimes.

One way to determine whether a dog has the potential for AAI is to administer a validated temperament test (see Table 10.1). These exist for both puppies and adult dogs, often for specific purposes, such as suitability for adoption, hunting, herding, and TD or service dog work. It is vital to seek the guidance of an expert in the area of specialization as part of the testing procedure (Davis, 2005). Additionally, because temperament can be affected by a dog's current health status, a veterinary examination prior to submitting a dog to a temperament test is recommended. A temperament test is akin to a snapshot in time, in that it gives a good evaluation of the dog's temperament on that day and in that situation. It is a good tool for detecting problem behaviors that need to be monitored or improved, but it gives limited information on what degree of improvement you might expect to see in those behaviors going forward. Dogs change as their life circumstances and experiences change, so giving multiple temperament tests to a dog living in the same stable environment for an extended period of time will be more revealing than a single test given to a dog living in a more temporary or short-term environment, or to a younger dog that has had fewer experiences.

Training the Dog for AAI Work

Selecting a dog with the appropriate temperament for AAI work is only the first step because training and experience can have a profound effect on behavior (McConnell & Fine, 2010). It is crucial to socialize a puppy and to continue to socialize an adult dog to a variety of different situations, people, and animals. It is the trainer/handler's responsibility to help the dog to find these varied environments rewarding and enjoyable, and the most straightforward way to accomplish this is through the use of positive reinforcement based on operant contingencies—that is, by giving the dog something he finds rewarding (e.g., a food treat or verbal praise) in those various situations, especially when the dog is in the act of performing a desired behavior (e.g., sitting politely for petting). Dog training is highly popular and visible in contemporary society, with a wealth of training resources, approaches, and formats (from books and DVDs to online and face-to-face classes). Generally, the evidence now indicates that positive reinforcement methods are more effective in the long term and demonstrate greater concern for animal welfare issues than those that rely on any form of punishment for undesired behaviors (Hiby, Rooney, &

Bradshaw, 2004; Tynes, 2011). See the Training Resources and TD Certifying Agencies sections of Table 10.1 for more information.

Well-trained dogs are much more predictable than poorly trained dogs, and the process of training the dog typically brings the handler a better understanding of the dog's strengths, weaknesses, and individual needs. Additionally, this process presumably brings the dog a greater understanding of what will be requested of him and a better ability to predict or interpret the handler's behaviors. Although we are focusing on the animal's perspective, it is important to remember that all AAI work is done in teams, typically dog and handler pairs, and that ultimate responsibility for the safety and well-being of the dog and the people with whom the dog interacts falls on the shoulders of the human part of the team. Training, experienced teamwork, and advanced preparation all put the handler in a better position to make good decisions.

Evaluating the Effect of AAI Work on the Dog

Once the dog is involved with AAI, it is important to evaluate whether the experience is rewarding or fun for him. But how can we tell if a dog is enjoying an experience or is stressed or uncomfortable? Because dogs cannot communicate verbally with us, we are forced to interpret their mental state from their behaviors, which can vary in subtle ways. AAI handlers must recognize differences between their dog tolerating an experience and enjoying it. To make this distinction and to evaluate the dog's genuine response to the "work," handlers must learn how to read the subtleties of their dog's behaviors. The section in Table 10.1 on Understanding Dog Behavior includes a number of useful resources to help with this. Following is a quick summary of some of the signs discussed in those sources.

Behavioral Indicators in AAI Dogs

Signs of Stress

- Panting (particularly fast or heavy panting) when it is not hot or the dog has not been physically active can be a sign of moderate to severe stress or illness.
- Pacing can be a sign of moderate to severe stress or illness.
- Ears flattened or pinned back is typically a sign of fear but can also indicate aggression.
- "Hard eyes"—eyes that are large, round, and typically fixed on the stressor—mean that this dog is likely to take the offensive.
- "Look away"—in most breeds, purposeful movement of the head away from a source of threat or stress is intended to reduce tension

by signaling to the source of concern that they mean no harm. Some dogs (especially herding breeds) tend to avoid direct eye contact, so the handler must know whether the dog typically offers eye contact behavior.

- Lip licking is a sign of low-level anxiety. The tongue typically comes out the front of the mouth to do a very quick lick of the front of the lips.

Danger Signs!

In addition to behaviors commonly associated with a stressed or angry dog (e.g., show of teeth, lip curl, growl), handlers must watch for these behaviors and take immediate action:

- Freeze: The dog stops moving or the body motion slows considerably. The body is typically in a rigid posture, and the eyes form a hard stare. This is indicative of a prebite behavior. Use extreme caution in removing this dog from the situation, but do so immediately.
- Whale eye: The dog has done a "look away" behavior, redirecting the head away from the source of stress (and may have even frozen in place), but now moves only its eyes in the direction of the stressor/threat. This is called whale eye because the eye is large and round, and when it happens the white part of the eye is sometimes visible on one side, making the eye appear like that of a whale. This dog is also very likely to bite and needs to be removed from the situation.

Displacement Behaviors

Things dogs may do to avoid stress:

- Yawning: Sometimes a yawn is just a yawn, but taken out of a sleep-related context a yawn is usually considered a sign of stress.
- Sniffing: It may be that there are simply interesting smells around, but if the dog has already had ample opportunity to explore this environment it is likely that the sniffing is a form of avoidance behavior. The handler must evaluate this behavior in context.
- Ignoring well-known cues: When the dog performs a routine behavior like "sit" readily and repeatedly on command, in a wide variety of contexts, and then suddenly doesn't sit when asked, it may be an indication of stress.

- The mouth is slightly open, eyes slightly squinted, body relaxed (not stiff or tensed) and flexible.
- The base of the ears is set midrange (forward ear set usually indicates alert behavior, whereas an ear set further back can mean appeasement or fear).

Tail Wag

The tail wag requires special consideration because people often mistakenly assume that any dog wagging its tail is relaxed and happy, but that is not necessarily the case. It is important to look at the dog's whole body to better interpret the tail wag. If the dog's mouth is slightly open, the eyes are slightly squinted, the base of the ears are set midrange, and the body is relaxed and moving flexibly with the tail wag, then all indications are that the dog is in fact relaxed and happy. On the other hand, if the dog's mouth is closed, the eyes are hard, the body is stiff, or the ears are pinned back, a wagging tail is not likely to be indicative of a relaxed or happy dog.

How to Reduce Stress for AAI Dogs

Play

It is hard to underestimate the importance of play for dogs because it influences communication, development, emotions, motivation, and behavior (London & McConnell, 2008). A wonderful means of building or improving a relationship between dog and handler, it is one of the best ways to reduce stress because play behaviors tend to be incompatible with stress behaviors. Every dog is different in the combinations of play it relishes, and therefore it is important to find ways that allow both the human and the dog to genuinely enjoy the play sessions (London & McConnell, 2008).

Exercise

It is important to make sure the dog has ample time to exercise before and after each AAI session. In dog training books or manuals, the word *exercise* is often used in place of the word *elimination*, presumably because it sounds more tasteful than referring directly to urination and excretion. However, in the AAI context, we need to make a clear distinction between these two concepts. Here the word *exercise* is used in a more general or traditional sense in that the dog gets to move its body by going for a walk, trot, or run, or by exploring the area. Obviously it is also important to make sure that the dog has an opportunity to relieve itself both before and after an AAI session, but that is not a sufficient "release." Letting dogs be themselves with no pressure to interact with another human (save the handler) or animal allows them to relax prior to the AAI session and helps them to decompress after the session. This relaxation time can include play, but whether it does should be the dog's choice. If the dog drops the Frisbee to sniff or to run after a leaf blowing in the wind, the handler should accept that behavior.

"Work" Schedule

It is the handler's responsibility to know his or her dog and to work well within the dog's individual boundaries for time spent in AAI sessions. The handler must take the dog's abilities, age, current and changing health status, and environmental stressors into account during each individual session and be prepared to end or cancel the session when appropriate. For example, as TDs age, adjustments must be made to take their welfare into account (Serpell, Coppinger, Fine, & Peralta, 2010). Older dogs may have less energy or reduced tolerance for certain environmental stressors, or they may begin to experience health issues that cause them discomfort. The handler may initially opt to reduce an older dog's participation in AAI sessions, and in some cases it may be best to stop the AAI participation completely.

Environmental Stressors

Handlers must be aware of and make an attempt to ameliorate environmental conditions that may stress the dog. When it is not possible to reduce or eliminate the environmental stressor, the handler must keep a careful watch over the dog and consider reducing the length of the session or ending the session completely. Following is a noninclusive list of examples:

- Sound, particularly loud, sharp, or high-pitched sounds: Children's vocalizations in particular can be alarming and stressful even for the most tolerant of dogs. Limit the dog's exposure to these noises and encourage children to speak quietly in the presence of the dog.
- Space: Like people, dogs possess an intimate or personal space (Davis, 2002) and can become stressed when others, particularly large numbers of people, enter that space. For example, a horde of enthusiastic children crowding around a dog to offer hugs and kisses can be highly stressful to the dog. There are a number of positive alternate behaviors that can be taught to children in this situation, but the handler's immediate priority is to protect the dog from being overwhelmed by a large group of well-meaning individuals. Related to this issue of space is hugging.
- Hugs: Although people hug to show affection, dogs do not. A dog may feel trapped or stressed when being hugged or may

simply not like the invasion of personal space. Most dogs can be trained to tolerate hugs, but again, tolerating is not the same as enjoying, and not all dogs are trained in this way. AAIs should not inadvertently teach children that hugging a dog is always acceptable. Ideally, an AAI session should teach children appropriate ways to interact with all dogs, not just TDs.

- Face-to-face interactions or kisses: Direct eye contact can be perceived by a dog as a threatening gesture, and as we pointed out earlier, some breeds are particularly uncomfortable with direct eye contact and may react very strongly to this sort of face-to-face interaction. Again, some dogs can be trained to tolerate this behavior, but because not all dogs are trained this way, it is important to teach people, particularly children, to avoid this behavior with the TD and other dogs they encounter.

WORKING WITH SPECIAL POPULATIONS

AAI with special populations brings a number of unique challenges and rewards. Here we focus on the relevant animal welfare concerns. For example, developmentally delayed or very young children, and physically or mentally disabled individuals of any age, may not be aware of actions that could inadvertently cause harm to animals. Children may not understand that dogs should not eat crayons or that putting a Lego brick into the dog's ear can harm the dog. Professional staff may opt to use these situations as learning opportunities for the population in question, but in all cases the handler must remain vigilant and intervene when necessary. Handlers must also avoid working with individuals they suspect have the potential to cause intentional harm to the TD. If the handler does not have a sense of confidence in the population he or she has been asked to work with, or in the staff in charge of monitoring that population, it is best to avoid bringing a TD into that environment.

It is important to visit the facility without the dog, meet the staff, and see the environment before bringing the dog. This allows the handler to ask a variety of questions and to establish rapport with staff in advance of the AAI session. What are the specific rules of the facility? What expectations should the handler have of the people and the environment? What do they expect of the handler and the dog? What procedures should be followed in the event of an emergency? Are there any objects or chemicals in the environment that may be harmful to the TD? It is important to ask specific questions; a more general question such as, "Is there anything in the room that might harm my dog?" will likely generate a "No" response. People often assume that if an environment is safe for children (or other special populations), it must also be safe for a dog, but that is not necessarily the case. Asking specifically whether the children play with balloons or toys with very small parts, or whether there will be any chocolate in the room, is more likely to generate a meaningful response, which is important because each of these examples represent a health risk to the TD if ingested.

Adult or child special populations can present causes for concern beyond potential ingestion hazards. Special populations are often located in environments containing specialized, unique equipment. The equipment may move (e.g., electric wheelchairs, chair lifts, bed elevators), make odd noises (e.g., various electronic monitors, public address systems, breathing assistance devices) or emit a variety of odors (from chemical/medicinal to biological in origin). All of these things have the potential to distract or alarm a dog. The best approach is to train the dog to be comfortable with these sorts of stimuli and perhaps even find them rewarding. Even then, the handler must be patient and calm to provide reassurance to the dog while at the same time being alert for circumstances that require action.

Finally, although infection control must be considered in any environment, it may be a more serious concern in specific settings. For example, in hospitals or rehabilitation centers bacterial infections, such as *Staphylococcus* and especially methicillin resistant *Staphylococcus aureus* (MRSA), can be contracted and spread by both the TD and the handler. TD certifying organizations are now helping to educate handlers on the topic of infection control, but again it is the responsibility of the handler to be informed about any potential risks associated with each facility visited and the specific steps that should be taken to minimize those risks. The individual TD-certifying agency and the specific facilities will have more detailed information on this topic.

VETERINARY CARE

Routine veterinary care is critical to ensuring the health, welfare, and well-being of all animals. It is recommended that dogs be seen twice a year by a licensed veterinarian and given a full physical examination. Veterinarians can provide guidance on appropriate preventive care; administer vaccines, parasite preventives, proper nutrition, and oral care; and address any abnormalities (e.g., lameness, increasing body weight, endocrine and organ dysfunction) or behavioral changes (e.g., lethargy, weakness, aggression, or indication of pain) in a timely manner (http://www.banfield.com/Pet-Owners/Our-Hospitals/Services/Preventive-Care).

CONCLUSION

Ensuring the welfare and well-being of animals in our care is not an option; it is an essential component to the success of the partnership between investigator and subject, human and dog. It is these additional considerations and activities described in this chapter that can make the difference between a life worth living and a life worth avoiding while in pursuit of our objective: understanding the neurophysiological bases of the relationship between animals and humans.

REFERENCES

Aloff, B. (2005). *Canine body language: A photographic guide*. Wenatchee, WA: Dogwise.

Americans With Disabilities Act of 1990, Pub. L. No. 101-336, § 2, 104 Stat. 328 (1991).

Barnard, S., Siracusa, C., Reisner, I., Valsecchi, P., & Serpell, J. A. (2012). Validity of model devices used to assess canine temperament in behavioral tests. *Applied Animal Behaviour Science*, *138*, 79–87. http://dx.doi.org/10.1016/j.applanim.2012.02.017

Berns, G. S., Brooks, A. M., & Spivak, M. (2012). Functional MRI in awake unrestrained dogs. *PLoS ONE*, *7*(5), e38027. http://dx.doi.org/10.1371/journal.pone.0038027

Brambell, F. W. R., & The Technical Committee. (1965). *Report of the technical committee to enquire into the welfare of animals kept under intensive livestock husbandry systems* (Vol. 2836 of Command Papers). London, England: Her Majesty's Stationary Office.

Broom, D. M. (1986). Indicators of poor welfare. *British Veterinary Journal*, *142*, 524–526. http://dx.doi.org/10.1016/0007-1935(86)90109-0

Broom, D. M. (1988). The scientific assessment of animal welfare. *Applied Animal Behaviour Science*, *20*, 5–19. http://dx.doi.org/10.1016/0168-1591(88)90122-0

Butler, K. (2004). *Therapy dogs today: Their gifts, our obligation*. Norman, OK: Funpuddle.

Coren, S. (2000). *How to speak dog: Mastering the art of dog-human communication*. New York, NY: Free Press.

Davis, K. D. (2002). *Therapy dogs: Training your dog to reach others* (2nd ed.). Wenatchee, WA: Dogwise.

Davis, K. D. (2005). *The canine behavior series: Temperament testing adult dogs*. Retrieved from http://www.veterinarypartner.com/Content.plx?P=A&A=2152

Donaldson, J. (1996). *The culture clash: A revolutionary new way of understanding the relationship between humans and domestic dogs*. Berkeley, CA: James & Kenneth.

European Parliament and Council of the European Union. (2010). EU Directive 2010/63/EU of the European Parliament and of the Council of 22 September 2010 on the Protection of Animals Used for Scientific Purposes. *Official Journal of the European Union, 276*, 33–79.

Farm Animal Welfare Council. (2009). *Farm animal welfare in Great Britain: Past, present and future*. London, England: Author.

Hiby, E. F., Rooney, N. J., & Bradshaw, J. W. S. (2004). Dog training methods: Their use, effectiveness and interaction with behavior and welfare. *Animal Welfare, 13*, 63–69.

Hubrecht, R. C., Serpell, J. A., & Poole, T. B. (1992). Correlates of pen size and housing conditions on the behavior of kenneled dogs. *Applied Animal Behaviour Science, 34*, 365–383. http://dx.doi.org/10.1016/S0168-1591(05)80096-6

Jones, A. C., & Gosling, S. D. (2005). Temperament and personality in dogs (*Canis familiaris*): A review and evaluation of past research. *Applied Animal Behaviour Science, 95*, 1–53. http://dx.doi.org/10.1016/j.applanim.2005.04.008

Kalnajs, S. (2007). *The language of dogs: Understanding canine body language and other communication signals* [DVD]. Madison, WI: Blue Dog Training & Behavior, LLC. Retrieved from http://www.bluedogtraining.com

Kruger, K. A., & Serpell, J. A. (2010). Animal-assisted interventions in mental health: Definitions and theoretical foundations. In A. H. Fine (Ed.), *Handbook on animal assisted therapy: Theoretical foundations and guidelines for practice* (pp. 33–48). San Diego, CA: Academic Press. http://dx.doi.org/10.1016/B978-0-12-381453-1.10003-0

London, K. B., & McConnell, P. B. (2008). *Play together, stay together: Happy healthy play between people and dogs*. Black Earth, WI: McConnell.

Loveridge, G. G. (1998). Environmentally enriched dog housing. *Applied Animal Behaviour Science, 59*, 101–113. http://dx.doi.org/10.1016/S0168-1591(98)00125-7

McCardle, P., McCune, S., Griffin, J., & Esposito, L. (2011). *Animals in our lives: Human-animal interaction in family, community and therapeutic settings*. Baltimore, MD: Brookes.

McConnell, P. B. (2006). *For the love of a dog: Understanding emotion in you and your best friend*. New York, NY: Ballantine Books.

McConnell, P. B. (2012). *Lending a helping paw: A guide to animal assisted therapy* [DVD]. Black Earth, WI: McConnell. Retrieved from http://www.patriciamcconnell.com

McConnell, P. B., & Fine, A. H. (2010). Understanding the other end of the leash: What therapists need to understand about their co-therapists. In A. H. Fine (Ed.), *Handbook on animal assisted therapy: Theoretical foundations and guidelines for practice* (pp. 149–165). San Diego, CA: Academic Press. http://dx.doi.org/10.1016/B978-0-12-381453-1.10009-1

Moesta, A., McCune, S., Deacon, L., & Kruger, K. (2015). Canine enrichment. In E. Weiss, S. Zawistowski, & H. Mohan-Gibbons (Eds.), *Animal behavior for shelter veterinarians and staff* (pp. 160–171). Hoboken, NJ: Wiley.

Pullen, A. J., Merrill, R. J. N., & Bradshaw, J. W. S. (2010). Preferences for toy types and presentations in kennel housed dogs. *Applied Animal Behaviour Science, 125,* 151–156. http://dx.doi.org/10.1016/j.applanim.2010.04.004

Pullen, A. J., Merrill, R. J. N., & Bradshaw, J. W. S. (2012a). The effect of familiarity on behavior of kennel housed dogs during interactions with humans. *Applied Animal Behaviour Science, 137,* 66–73. http://dx.doi.org/10.1016/j.applanim.2011.12.009

Pullen, A. J., Merrill, R. J. N., & Bradshaw, J. W. S. (2012b). Habituation and dishabituation during object play in kennel-housed dogs. *Animal Cognition, 15,* 1143–1150. http://dx.doi.org/10.1007/s10071-012-0538-2

Rawlings, J., Deacon, L., & Healey, L. (2007). *A non-invasive method of collecting saliva and plaque in dogs and cats* (NC3Rs Parliamentary Showcase). Retrieved from http://www.nc3rs.org.uk/

Rooney, N., Gaines, S., & Hiby, E. (2009). A practitioner's guide to working dog welfare. *Journal of Veterinary Behavior: Clinical Applications and Research, 4,* 127–134. http://dx.doi.org/10.1016/j.jveb.2008.10.037

Russell, W. M. S., & Burch, R. L. (1959). *The principles of humane experimental technique.* Baltimore, MD: Johns Hopkins Bloomberg School of Public Health. Retrieved from http://altweb.jhsph.edu/pubs/books/humane_exp/het-toc

Sandoe, P., & Christiansen, S. B. (2008). *Ethics of animal use.* Chichester, West Sussex, England: Wiley-Blackwell.

Serpell, J. A., Coppinger, R., & Fine, A. H. (2006). Welfare considerations in therapy and assistance animals. In A. H. Fine (Ed.), *Handbook of animal assisted therapy: Theoretical foundations and guidelines for practice* (pp. 453–474). San Diego, CA: Academic Press. http://dx.doi.org/10.1016/B978-012369484-3/50021-9

Serpell, J. A., Coppinger, R., Fine, A. H., & Peralta, J. M. (2010). Welfare considerations in therapy and assistance animals. In A. H. Fine (Ed.), *Handbook on animal assisted therapy: Theoretical foundations and guidelines for practice* (pp. 481–503). San Diego, CA: Academic Press. http://dx.doi.org/10.1016/B978-0-12-381453-1.10023-6

Turid, R. (2005). *On talking terms with dogs: Calming signals.* Rugeley, England: Qanuk.

Tynes, V. V. (2011). *Why punishment fails; what works better.* Retrieved from http://files.dvm360.com/alfresco_images/DVM360/2013/11/11/81223004-aefb-49be-9895-0cc3f8e2abc6/article-735183.pdf

United Kingdom Home Office. (2012). *Consultation on options for the transposition of European Directive 2010/63/EU on the protection of animals used for scientific purposes: Summary report and government response.* Retrieved from https://www.gov.uk/government/uploads/system/uploads/attachment_data/file/157919/summary-response-transposition.pdf

U.S. Department of Agriculture. (2013). *Animal Welfare Act and animal welfare regulations.* Retrieved from http://www.aphis.usda.gov/animal_welfare/downloads/Animal%20Care%20Blue%20Book%20-%202013%20-%20FINAL.pdf

U.S. Department of Justice. (2010). *Service animals: ADA requirements*. Retrieved from http://www.ada.gov/service_animals_2010.htm

Watson, A., Fray, T., Clarke, S., Yates, D., & Markwell, P. (2002). Reliable use of the ServoMed Evaporimeter EP-2 to assess transepidermal water loss in the canine. *The Journal of Nutrition, 132*(6, Suppl. 2), 1661S–1664S.

Wells, D. L. (2004). A review of environmental enrichment of kenneled dogs, *Canis familiaris*. *Applied Animal Behaviour Science, 85*, 307–317. http://dx.doi.org/10.1016/j.applanim.2003.11.005

Yeates, J. W. (2011). Is "a life worth living" a concept worth having? *Animal Welfare, 20*, 397–406.

INTEGRATIVE COMMENTARY III: A PRIMER IN THREE AREAS KEY TO FUTURE RESEARCH

PEGGY McCARDLE

Part III of this volume, Science and Research Considerations, comprises three chapters addressing areas that can and should play a key role in future research in human–animal Interaction (HAI). Certainly the first two parts of this volume—addressing behavior and neuroscience as they might lead to focused research on the social neuroscience of HAI—are areas that must be considered as we study the underlying mechanisms of the interaction between humans and animals, especially as they influence the inextricable link between social-emotional and neurological development and health. The three chapters in Part III offer fundamental information on genetics, salivary bioscience, and animal welfare, three areas that have much to offer as we move forward in this research.

Most of us know that the human genome has been mapped, and we may be peripherally aware that the genetics of dogs and cats are also topics of intense study. We may have seen the cover of *Science* for the issue that announced the

http://dx.doi.org/10.1037/14856-014

The Social Neuroscience of Human–Animal Interaction, L. S. Freund, S. McCune, L. Esposito, N. R. Gee, and P. McCardle (Editors)

mapping of the dog genome (Sutter et al., 2007), showing a Chihuahua and a Great Dane, two dramatically different examples of dogs. Jones and McCune (Chapter 8) introduce us to how animal genetics has been and might in future be highly useful in the context of HAI research and practice. Genetic information that can guide us to the selection of the most behaviorally well-suited dogs for animal-assisted interventions is just one obvious example. But Jones and McCune go deeper, into the "breed-cluster and breed-specific behaviors and morphological and neurochemical genetic drivers of behavior" (p. 178). They discuss the origins of certain breed-specific behaviors, explaining how domestication and genetic selection have led to behaviors that have over time resulted in stronger dependencies between humans and animals. As DNA technology changes rapidly, these authors predict equally rapid progress in behavioral genetics research on companion animals. This work should help us better understand what drives behavior, trainability, and optimal pet and therapy animal selection. Jones and McCune offer optimism as to the possibility of more tailored HAIs, as well as genetics research that will enable us to delve more deeply into the mechanisms that underlie the benefits that have been and continue to be documented for the interaction between humans and animals, in the home, the clinic, and the research laboratory.

Given the general focus on HAI research that has documented reductions in stress and loneliness, and improvements in health that have been linked to humans' involvement with pets and companion animals, the underlying science of stress-related and other hormones is clearly a path forward in seeking to understand mechanisms. Dreschel and Granger (Chapter 9) offer us a literal "how to" guide for the inclusion of salivary measures in HAI research. (Note that work on the role of other hormones such as oxytocin and vasopressin is addressed in Chapters 4 and 5 in Part II, by Carter & Porges and Beetz & Bales, respectively.) Because salivary measures, compared with blood sampling, are less invasive for both the human and the animal, they have the potential dual benefit of providing a low-stress view into the physiological impact of HAI while meeting higher animal welfare standards. Additionally, salivary measures can be collected repeatedly and with greater temporal precision than other peripheral measures such as urine samples. Dreschel and Granger, in Chapter 9, provide us with primers on both human and animal salivary bioscience. They address sample collection, possible contamination, sample handling, transport, and storage, and even research designs, how and when to sample, and guidelines on analysis. The fact that this is a rapidly advancing science with ever more substances being reliably measured in saliva gives salivary bioscience great potential for use in HAI research.

Animal welfare should seem an obvious factor in any research that involves animals, yet much of the focus of HAI research has been on the human side of the interaction: How does interaction with animals help humans stay healthier, cope with disability, be less stressed or more motivated, etc.? What must be remembered is that both human research participant protections and animal welfare assurances should be in place for HAI research. Indeed, the first NIH Requests for Applications addressing HAI called for protection certifications for both animals and humans (RFA-HD-09-030, http://grants. nih.gov/grants/guide/rfa-files/RFA-HD-09-030.html; and RFA-HD-09-031, http://grants.nih.gov/grants/guide/rfa-files/RFA-HD-09-031.html). As we move to more intensive study of the animals involved in HAI, as foreshadowed by Jones and McCune in Chapter 8, toward greater collection and use of animal saliva (with its challenges in sample collection) as outlined by Dreschel and Granger, in for example, stress response, animal welfare is a key consideration. In addition, as we seek to attract more researchers from various disciplines who might not have previously involved animals in their work, it becomes ever more important that all researchers have heightened awareness of how best to safeguard animal welfare.

Gee, Hurley, and Rawlings (Chapter 10) not only offer us a timely reminder that there are two sides to the interaction between humans and animals that must be considered, but also guide us beyond the required minimum, calling HAI research to a higher standard. They point out that with such a higher standard, everyone benefits—the animals are healthier and happier, learn faster, and produce better results. This chapter defines animal welfare, guides the HAI researcher in providing for it, and—using dogs as the example—defines what we can expect of dog behavior in the practice setting and how to select dogs for therapeutic work. The authors address training (which is important not only in therapeutic and home-environment interactions, but also in the least-stressful approach to blood and saliva sample collection) and how to reduce stress in the dog. They provide us with the basics and important resources on behavioral indicators, a crucial area for all those working with animals or owning pets, and address specific issues that arise in involving companion and therapy dogs with special human populations. There is much to learn from this chapter about how best to work with dogs (and animals in general) in the context of both HAI research and practice.

As we seek to better understand how and why genetics is important to HAI, the potential for salivary bioscience in studying HAI, and how best to care for and protect our animal research (and animal-assisted intervention) participants just as we do our human research participants, these three chapters provide information that is invaluable to every researcher and clinician. Whether the goal is simply to have a better grasp of the research literature, or to include these techniques in one's own research (although clearly this will require the building of an interdisciplinary team with individuals who have sufficient expertise in any of these areas), the authors offer us a valuable education. These chapters should serve as a useful guide to the very real possibilities that lie ahead for HAI research as we integrate it with social neuroscience.

REFERENCE

Sutter, N. B., Bustamante, C. D., Chase, K., Gray, M. M., Zhao, K., Zhu, L., . . . Ostrander, E. A. (2007). A single *IGF1* allele is a major determinant of small size in dogs. *Science, 316,* 112–115. http://dx.doi.org/10.1126/science.1137045

IV

CONCLUSION

FINAL COMMENTARY: SOCIALITY, THERAPY, AND MECHANISMS OF ACTION

NATHAN A. FOX

I want to be the kind of person my dog thinks I am.

—Author unknown

The chapters in this interesting volume address a number of important issues with regard to human–animal interaction (HAI). Three areas in particular stand out: One is the issue of sociality, the idea that across evolution many species have evolved highly specialized behaviors that facilitate social interaction among conspecifics. These behaviors provide multiple functions, including protection from threat and community food gathering. In addition, at least in the human condition, social behaviors provide the foundation for the establishment of an attachment relationship between an infant and caregivers early in life and form the basis for subsequent social relationships across development. Animals, particularly domesticated ones such as dogs or cats, are part of that circle of social relationships that humans have established. As such, these animals can serve an important function in providing social comfort within a larger network of relationships. A second area addressed in these chapters is the manner in which animals may serve an important therapeutic

http://dx.doi.org/10.1037/14856-015
The Social Neuroscience of Human–Animal Interaction, L. S. Freund, S. McCune, L. Esposito, N. R. Gee, and P. McCardle (Editors)

or educational function. There are now many studies that show the benefits of animals as mediators of therapeutic assistance to people who are undergoing painful surgical procedures or are involved in therapy for psychological distress. The presence of an animal, usually a dog, appears to reduce stress and facilitate positive affect as well as promote positive health benefits for individuals. The third area covered in the chapters in this volume has to do with mechanisms of action. Many of the chapters discuss the physiological systems that are triggered by the presence of a domesticated animal or pet and how activation of these systems reduces stress and enhances positive outcomes. Much of the discussion in these chapters is whether these mechanisms of action are bidirectional, that is, whether they occur not only in the human recipient but in the animal as well. In addition, the mechanisms by which these physiological systems may buffer stress is considered.

SOCIALITY

In Chapter 6, Brown and Coan present an interesting argument about the way in which social proximity is in part responsible for the manner in which humans regulate their emotions. They outline a model in which social relationships are perceived as resources for accomplishing goals. By asking the question as to why humans put so much energy into social interactions, they provide a context for a theory of social affect regulation in which social relationships protect against the negative effects of stress and may actually be protective against the demands of a threatening environment. They present what they call the *social baseline theory*, which argues that most human emotion regulation takes place within an environment or network of relationships. In this case, the "baseline" situation for humans is the social network, and isolation is the absence of these supports. In their chapter, they write "social relationships, cooperative behavior, and interdependence have benefitted human energy use and allocation over evolutionary time such that humans could evolve to fit a social ecology, as opposed to a specific physical ecology" (p. 133). Brown and Coan argue that animals (in particular, dogs) provide companionship and social support much as other humans do and that they are part of the social network of relationships that provides important affect regulation benefits to humans. Their social baseline theory is reminiscent of work by Myron Hofer, a psychiatrist and psychobiologist who in 1984 published a paper titled "Relationships as Regulators: A Psychobiologic Perspective on Bereavement" in which he discussed the importance of social relationships, including initial attachment relationships between infant and caregiver, as having prime importance in regulating an individual's affect. Relationships are the regulators of emotion, much as they are in Brown and

Coan's social baseline model. Hofer cited examples of the psychological and biological consequences among individuals who have lost their significant partners. Changes in their mood and affect parallel changes in autonomic and hormonal reactivity and often result in depression, disoriented cognition, and autonomic instability. These changes are similar to those seen in individuals who have undergone sensory deprivation and isolation. Hofer looked to the developmental origins of the need for social affiliation in the emergence of attachment relationships between infant and caregivers.

Themes related to the importance of sociality and attachment and social relationships also appear in the chapters by Carter and Porges (Chapter 4) and by Beetz and Bales (Chapter 5). In both, there is an emphasis on how social interactions have particular value in reducing stress, enhancing feelings of positive well-being, and providing opportunities for emotion regulation. Both chapters provide a context for the emergence of sociality in the emergence of affiliative and caregiving behaviors in animals, and both provide an overview of the physiological systems that appear to underlie these behaviors. In both chapters, there is an emphasis on the evolution of function of the autonomic nervous system (with an emphasis on Porges's polyvagal theory) and on the role of oxytocin in enhancing social relationships. Two aspects of the polyvagal theory are important here: The first has to do with the evolution of the neural systems to produce and receive complex acoustic and facial cues associated with social communication, and the second is the role of the more recently evolved myelinated vagal system that physiologically supports oxygenation of the neocortex and is necessary for social engagement, leading to affective states of safety and security.

Oxytocin also plays an important role in reducing stress and promoting feeling states of safety and security. Carter and Porges argue that oxytocin may bias autonomic responses and may be involved in down-regulation of the HPA axis, the main stress system in the body. Beetz and Bales write about the role that oxytocin plays in maternal–infant bonding as well as in parental behavior. They cite studies of the sociogenic effects of oxytocin administration on human behavior, including increased eye contact, trust, and positive self-perception and enhanced social skills. Together, these chapters make a strong argument for understanding the physiological bases of social behavior and the manner in which these systems have evolved across species to provide a foundation for basic sociality.

Another perspective on sociality is presented in the chapters by Paul Quinn (Chapter 1) and Kun Guo (Chapter 2). Quinn asks the question as to what infants know about cats, dogs, and people. Using the tools of experimental infant psychology, Quinn examines the perceptual and conceptual features that infants use to discriminate between different categories of animals. As Quinn states, understanding infants' categorization of nonhuman animals is significant because it provides insight into the human ability to divide the world of objects into perceptual categories and to move from there to more conceptual distinctions. It also provides understanding of how we humans begin to view nonhuman animals and hence how we then go on to treat them in our social world. One of the insights of this work is the research finding that infants come to include nonhuman animals into the representation of humans. As Quinn writes, "humans may be the glue that provides the coherence for the concept of animal" (p. 25). It seems that a message from this work is that infants develop the knowledge that nonhuman animals, particularly those in their local surroundings, are part of their social network of relationships. This speaks to the notion of sociality and the incorporation of animals into the human social world.

The chapter by Guo conveys a similar message. Guo reviews what we know about facial recognition by both humans and nonhumans. This chapter reviews the literature on human face identification and nonhuman primate face identification findings, with striking parallels. And it presents interesting data on the use of gaze-direction information that is used by both humans and nonhuman primates to signal social intention. The point here is that both humans and animals extract important information about conspecifics from the face, and they use this information to interpret their social intentions.

HAI AS A THERAPEUTIC AND EDUCATIONAL CONTEXT

Many of the chapters in this volume comment on studies that have shown the benefits of HAI within medical and psychiatric contexts. The presence of animals, particularly dogs, appears to have a calming and stress-reducing effect, with changes in autonomic indices (heart rate, blood pressure) being the most common metrics. In addition, there is the interesting idea, in the chapter by Ling, Kelly, and Diamond (Chapter 3), that interaction with animals both within the family and the classroom may actually enhance learning and facilitate the development of executive functions. These authors speculate that having a pet in the family can be viewed as an opportunity to exercise and develop the important cognitive skills of inhibitory control, attention shifting, and planning. For example, these skills are known to improve with training, repeated practice, and incremental demands, something that caring for a pet in a household would do for a child within the network of the family relationship. The routines and responsibilities of pet ownership instill a sense of control and enhance important cognitive skills. Lozier, Brethel-Haurwitz, and Marsh, in Chapter 7, address the possibility of human–animal interaction providing the context for the learning of skills associated with empathy. Having to deal with a pet and provide food, water, and play creates opportunities for engaging in non-threatening social interaction as well as heightened awareness of the need for care, all of which together may enhance empathic skills. The emotional engagement that is elicited by HAI is an important context for learning of social skills.

An important coda to the work on HAI is contained in the chapter by Gee, Hurley, and Rawlings (Chapter 10). They address the issue of the use of animals for research and exactly what are the "rights" of animals. They distinguish between animals that are "used" for service or in therapeutic settings versus those kept as pets and those who are the subjects of experimentation. In their chapter, they provide a guide for identifying which types of animals might be best for particular therapeutic contexts. They discuss the issues of individual differences or temperament as important aspects of animal behavior. As a side note, the study of temperament has a rich history in animal research, particularly with dogs, but also with other species. Humans have selected and bred animals (particularly dogs) for specific characteristics (hunting, retrieving, and protection), and these individual differences can be assessed early in life and are as well subject to change as a result of early life experiences (Scott & Fuller, 1965).

MECHANISMS FOR UNDERSTANDING HAI

The individual differences noted in the Gee, Hurley, and Rawlings chapter lend themselves nicely to the discussion by Jones and McCune in Chapter 8 of the genetics of companion animals. The years of inbreeding for specific physical and behavioral characteristics of both dogs and cats and the advances in methods for parsing the genome make the study of genetics in companion animals a fruitful and important avenue for understanding the origins of certain behaviors. Jones and McCune give a good overview of these genetic methods and provide an interesting commentary on the future of this avenue of research.

Across many of the chapters (especially Chapters 4 and 5) there is extended discussion of the physiology underlying the behavioral effects of HAI. Much of that work focuses on oxytocin, but there is also discussion of other peptides (vasopressin) and of stress hormones such as cortisol. In Chapter 9, the ability to monitor these physiological changes is outlined by Dreschel and Granger with what they call *salivary bioscience*. They provide an overview of the methods for extracting hormonal information from saliva and hold open the possibility that one can use these methods with animals to examine physiological changes as a result of HAI.

CODA

In some ways, the study of HAI has found multiple niches within the psychological community. There are now multiple studies of the therapeutic benefits of HAI across multiple contexts and conditions. And the underlying physiological mechanisms of why HAI has these benefits are just beginning to be understood. As interesting and important is the move from understanding social communication and cognition in nonhuman primates (monkeys of different species) to the study of dogs. Experimental studies of dogs have begun to describe the context sensitivity and use of social cues by domesticated dogs (Call, Bräuer, Kaminski, & Tomasello, 2003; Gácsi, Miklósi, Varga, Topál, & Csányi, 2004). These studies have much to tell us about human social cognition, and they are enriching our understanding of the benefits of HAI. The current set of chapters is an important landmark in the study of HAI and of the importance of HAI to human social communication and well-being.

REFERENCES

Call, J., Bräuer, J., Kaminski, J., & Tomasello, M. (2003). Domestic dogs (Canis familiaris) are sensitive to the attentional state of humans. Journal of Comparative Psychology, 117, 257–263. http://dx.doi.org/10.1037/0735-7036.117.3.257

Gácsi, M., Miklósi, Á., Varga, O., Topál, J., & Csányi, V. (2004). Are readers of our face readers of our minds? Dogs (Canis familiaris) show situation-dependent recognition of human's attention. Animal Cognition, 7, 144–153. http://dx.doi.org/10.1007/s10071-003-0205-8

Hofer, M. A. (1984). Relationships as regulators: A psychobiologic perspective on bereavement. Psychosomatic Medicine, 46, 183–197. http://dx.doi.org/10.1097/00006842-198405000-00001

Scott, J. P., & Fuller, J. L. (1965). Genetics and the social behavior of the dog. Chicago, IL: University of Chicago Press.

FUTURE RESEARCH:
NEEDS AND PROMISE

LAYLA ESPOSITO, NANCY R. GEE, LISA S. FREUND,
SANDRA McCUNE, AND PEGGY McCARDLE

As research continues to indicate the potential for human health benefits from interacting with animals, it is becoming increasingly important to be able to identify the underlying mechanisms that are responsible for these gains in health and well-being. Only through this understanding will scientists and practitioners be able to harness and hone the incredible power of social and therapeutic relationships between humans and animals to improve the lives of both. As the field of human–animal interaction (HAI) considers a number of theoretical explanations and potential underlying mechanisms, such as attachment theory, bonding, biophilia, social support, and stress reduction, it is clear that social neuroscience is one of the most promising fields through which to investigate the possible neurobiological processes underpinning these effects. Given the growth in social neuroscience, both in theory and methods of investigations (e.g., Harmon-Jones & Beer, 2009), now is an ideal time for building multidisciplinary collaborations between

http://dx.doi.org/10.1037/14856-016

The Social Neuroscience of Human–Animal Interaction, L. S. Freund, S. McCune, L. Esposito, N. R. Gee, and P. McCardle (Editors)

researchers interested in HAI and those studying the neurobiological foundations of human and animal behavior.

Each of the chapters in this volume provides ideas about future research directions that should be considered at the intersection of social neuroscience and HAI science. Many of these are reviewed in Fox's commentary and are also integrated into the research suggestions we present in this chapter. We hope that these will guide and encourage researchers across these several disciplines to initiate or continue research that delves into the mechanisms that underlie the potential benefits and those already documented for HAI and animal-assisted interventions (AAIs).

On the basis of discussions at the original NICHD/WALTHAM Social Neuroscience of HAI workshop, the material presented in this volume, and other inputs, we highlight next a number of potential areas for future investigation:

- Investigation of whether processing advantages such as configural processing or attractor effects accrue for infants being reared with pets in the home, and whether, for example, category learning in infancy is one component of tuning perceptual systems.
- Systematic study of facial processing and communication using current methods such as eye tracking, electroencephalography, and electromyography to delve into the neurobiological underpinnings of the use of gaze and response to visual cues observed in HAI.
- Study of whether HAI may provide a naturalistic means of enhancing executive functions, reducing stress, building social skills, and improving learning and motivation in children and youth; such studies should examine short and longer term effects, and both direct and indirect effects.
- Neurobiologically, investigations should focus on neuropeptides, especially oxytocin and vasopressin, which play an important role in the integrated neural networks that coordinate sociality as well as other bodily processes; such research will be important not only to the understanding of the mechanisms that may underlie HAI effects, but also to the understanding of the evolutionary and neurobiological bases of mammalian society more generally.
- Other physiological measures should also be studied in conjunction with HAI, including salivary cortisol, heart rate and its variability, and electrophysiological measures (as noted earlier for studies of gaze, but also useful more generally in the study of HAI). Salivary cortisol is considered highly important

in studying stress and threat response, and should be studied (as should all these measures) in both humans and animals—that is, in *all* participants within HAI.

- Behavioral signs of stress, threat, aggression, and fear should be studied in greater depth, both in humans and animals, within various contexts and situations, and where possible linked to biological (neurophysiological) measures.

In addition to the suggestions for future research that have emerged from this volume, there are several overarching areas related to the social neuroscience of HAI that are in need of investigation.

As technologies in the biomedical and behavioral sciences are developed and used for increasingly diverse purposes, we are now in an era in which we can finally begin to explore more basic science related to the human–animal bond and therapeutic interactions with animals. Functional magnetic resonance imaging (fMRI), which can measure brain activity, appears to be a promising tool. An example of this type of groundbreaking research is a recent study by Stoeckel, Palley, Gollub, Niemi, and Evins (2014), which used fMRI to compare the neural responses of mothers viewing images of their child with those of the mothers viewing images of their own dog. The results indicated that there was significant overlap in brain activation in areas related to affiliation, emotion, reward, visual processing, and social cognition. There were also interesting differences in areas of activation that may also help us understand how bonds, attachments, and relationships between humans and pets vary. It is this type of work that has the potential to advance the field light years ahead of where it has been over the past several decades.

Dogs are now being trained to undergo fMRI scans, and though the work is still in the very early stages, we are in a unique position to catch a glimpse into real-time brain activity of our canine companions. For example, Andics, Gácsi, Faragó, Kis, and Miklósi (2014) presented dogs and humans with a set of vocal and nonvocal stimuli and were able to demonstrate that vocal areas exist in the brains of dogs and that the pattern of activation in those areas is similar to that seen in humans. Berns, Brooks, and Spivak (2015) presented dogs with scents of familiar and unfamiliar humans and dogs while they were undergoing fMRI scanning. The results demonstrated that the caudate was activated by the scent of the familiar human. This area of the brain has been implicated in positive expectations and social rewards, so one might conclude that the experience of smelling a familiar human is a positive one. Unfortunately, most brain regions have multiple functions, so we are not able to infer a particular cognitive state from an activation pattern like this one. Though these results are intriguing, researchers must take caution in drawing unwarranted conclusions and consider this work to be a helpful starting point for future investigations into the social neuroscience of HAI.

It is clear from the work presented here that there is still much to be learned about the human–animal bond itself. How are human–animal bonds different from human–human bonds (e.g., biologically, functionally, evolutionarily), and how are they different from animal–animal bonds? Are there unique benefits of human–animal bonds that cannot be gained from other types of interactions? For example, why do animals sometimes facilitate or elicit social interaction from individuals with certain social deficits (such as in autism spectrum disorder; e.g., O'Haire, McKenzie, Beck, & Slaughter, 2015)? We can apply social neuroscience to begin to address these questions.

The need for greater methodological rigor in HAI studies has been highlighted before (McCune et al., 2014). Today we are seeing more studies using appropriate design and statistical methods, but there is still cause for concern (Griffin, McCune, Maholmes, Hurley, 2011). An overview of methodological issues and strategies for building a stronger evidence base for AAIs are reviewed by Kazdin (2011).

The work presented in this volume summarizes the foundational starting point of the social-neuro-scientific investigation of human–animal interaction. The common theme running through all of the chapters herein is that there is much work to do. There are many new and developing technologies to exploit and theories to be tested. We urge researchers to take advantage of those new technologies, but to do so in a way that uses sound methodology, reliable and valid instrumentation, and standardized measures whenever possible (to allow for cross-study comparisons) and that considers both human and animal welfare and derives justified conclusions.

REFERENCES

Andics, A., Gácsi, M., Faragó, T., Kis, A., & Miklósi, Á. (2014). Voice-sensitive regions in the dog and human brain are revealed by comparative fMRI. *Current Biology, 24*, 574–578. http://dx.doi.org/10.1016/j.cub.2014.01.058

Berns, G. S., Brooks, A. M., & Spivak, M. (2015). Scent of the familiar: An fMRI study of canine brain responses to familiar and unfamiliar human and dog odors. *Behavioural Processes, 110*, 37–46. http://dx.doi.org/10.1016/j.beproc.2014.02.011

Griffin, J. A., McCune, S., Maholmes, V., & Hurley, K. (2011). Human-animal interaction research: An introduction to issues and topics. In P. McCardle, S. McCune, J. Griffin, & V. Maholmes (Eds.), *How animals affect us* (pp. 3–9). Washington, DC: American Psychological Association.

Harmon-Jones, E., & Beer, J. (Eds.). (2009). *Methods in social neuroscience*. New York, NY: Guilford Press.

Kazdin, A. (2011). Establishing the effectiveness of animal-assisted therapies: Methodological standards, issues and strategies. In P. McCardle, S. McCune, J. Griffin, & V. Maholmes (Eds.), *How animals affect us* (pp. 35–51). Washington, DC: American Psychological Association. http://dx.doi.org/10.1037/12301-002

McCune, S., Kruger, K. A., Griffin, J. A., Esposito, L., Freund, L. S., Hurley, K. J., & Bures, R. (2014). Evolution of research into the mutual benefits of human-animal interaction. *Animal frontiers*, *4*, 49–58. http://dx.doi.org/10.2527/af.2014-0022

O'Haire, M. E., McKenzie, S. J., Beck, A. M., & Slaughter, V. (2015). Animals may act as social buffers: Skin conductance arousal in children with autism spectrum disorder in a social context. *Developmental Psychobiology*, *57*, 584–595. http://dx.doi.org/10.1002/dev.21310

Stoeckel, L. E., Palley, L. S., Gollub, R. L., Niemi, S. M., & Evins, A. E. (2014). Patterns of brain activation when mothers view their own child and dog: An fMRI study. *PLoS ONE*, *9*(10), e107205. http://dx.doi.org/10.1371/journal.pone.0107205

INDEX

ABOUT THE EDITORS

Lisa S. Freund, PhD, is the chief of the Child Development and Behavior Branch at the *Eunice Kennedy Shriver* National Institute of Child Health and Human Development (NICHD), U.S. National Institutes of Health. She is a developmental neuropsychologist who is known for her neuro-imaging studies with children from different clinical populations and was an NICHD-supported scientist for several years. In the past, she had a private clinical practice that included equine animal-assisted therapy. She is currently responsible for a multifaceted research and training program at the NICHD to promote investigations, both basic and applied, to gain a deeper understanding of the developing brain and associated behaviors. She has been involved with the Mars–NICHD public-private partnership since it was established, including participation in the development of the partnership's sponsored workshops and the edited volume on human–animal interaction (HAI), *Animals in Our Lives* (McCardle, McCune, Griffin, Esposito, & Freund, 2011). Dr. Freund continues to serve in a leadership role for the HAI research-funding program at the NICHD.

Sandra McCune, PhD, is scientific leader for HAI at the WALTHAM Centre for Pet Nutrition (part of Mars, Inc.) in the United Kingdom. Her background is in ethology, and she has studied a range of topics in cat and dog behavior and welfare for many years, including aspects of temperament, social behavior, feeding behavior, cognition, and age-related changes in behavior. Her doctoral study focused on the assessment of individual variation in the temperament of cats and its impact on their welfare when confined. She has extensive experience studying HAI from both animal and human perspectives in a variety of contexts. Dr. McCune was instrumental in the establishment of the Mars–NICHD public-private partnership on child development and HAI, in planning the workshops sponsored by that partnership, and in editing two previous volumes on HAI under the partnership: *How Animals Affect Us: Examining the Influence of Human–Animal Interaction on Child Development and Human Health* (McCardle, McCune, Griffin, & Maholmes, 2011) and *Animals in Our Lives: Human–Animal Interaction in Family, Community, and Therapeutic Settings* (McCardle, McCune, Griffin, Esposito, & Freund, 2011). She continues to play a leadership role in the partnership.

Layla Esposito, PhD, is a program official within the Child Development and Behavior Branch, NICHD, and oversees the portfolio of research on social and emotional development in children, child and family processes, and human–animal interaction. She is a child psychologist by training and has been involved in a wide variety of research projects related to child and adolescent development. Dr. Esposito, involved with the Mars–NICHD public-private partnership since its inception, has developed the research program on HAI that she now oversees and has contributed to the development of HAI-related workshops and publications, including the edited volumes *How Animals Affect Us: Examining the Influence of Human–Animal Interaction on Child Development and Human Health* and *Animals in Our Lives: Human–Animal Interaction in Family, Community, and Therapeutic Settings*.

Nancy R. Gee, PhD, is research manager for the WALTHAM HAI research program. Based at the WALTHAM Centre for Pet Nutrition (part of Mars, Inc.) in the United Kingdom, she manages a global portfolio of external university collaborations. This role involves participation in the Mars–NICHD public-private partnership, development of other partnerships, and shepherding a new focus of research and practice for HAI at WALTHAM. Dr. Gee also holds the rank of professor of psychology at the State University of New York, Fredonia, where she has conducted research and published in the areas of cognition and HAI. A recipient of multiple grants and awards, she is a member of the editorial advisory boards for two journals, has served extensively as a reviewer of HAI research grant proposals, and has contributed chapters to this and other HAI volumes.

Peggy McCardle, PhD, MPH, is an affiliated research scientist at the Haskins Laboratories, New Haven, Connecticut, and an independent consultant. She is involved in editing volumes related to literacy and learning, mentoring young scholars and researchers, and consulting and writing in a variety of areas including child language development and learning, bilingualism, education, and learning disabilities. As former chief of the Child Development and Behavior Branch, she was closely involved with the establishment of the Mars–NICHD public-private partnership on child development and HAI, the workshops held under that partnership, and the editing of the two previous volumes, *How Animals Affect Us: Examining the Influence of Human–Animal Interaction on Child Development and Human Health* and *Animals in Our Lives: Human–Animal Interaction in Family, Community, and Therapeutic Settings.* Dr. McCardle continues to consult in HAI as it relates to human development, learning, and education.